I0464488

You and the Universe, a Book of Numbers

You and the Universe, a Book of Numbers

You and the Universe, a Book of Numbers

You and the Universe, a Book of Numbers

You and the Universe

A BOOK OF NUMBERS

BY

ERNEST C. WILSON

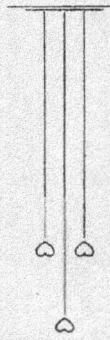

1922

The Harmonial Publishers

4328 Alabama Street,

San Diego, Calif.

You and the Universe, a Book of Numbers

TABLE OF CONTENTS

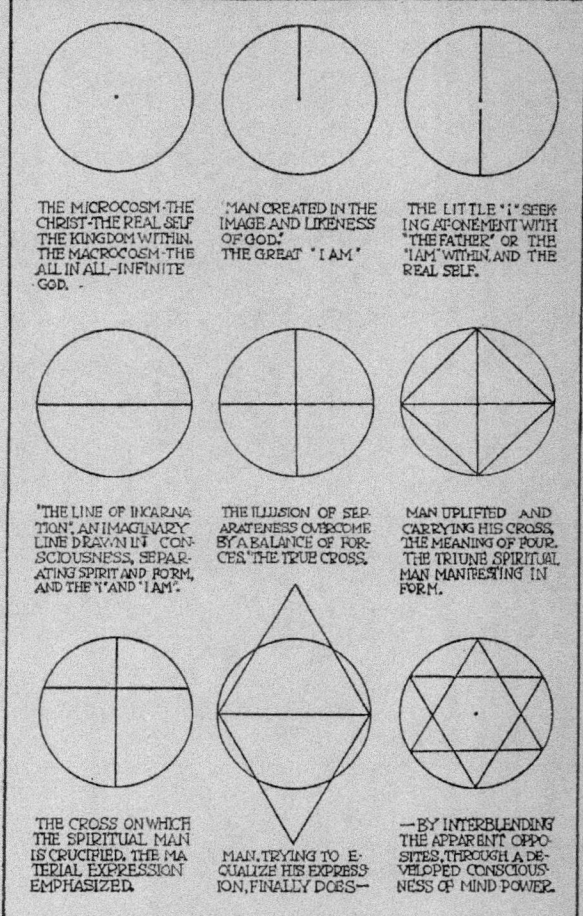

THE MICROCOSM-THE
CHRIST-THE REAL SELF
THE KINGDOM WITHIN.
THE MACROCOSM-THE
ALL IN ALL,-INFINITE
GOD.

MAN CREATED IN THE
IMAGE AND LIKENESS
OF GOD.
THE GREAT "I AM"

THE LITTLE "i" SEEK-
ING AT-ONEMENT WITH
"THE FATHER" OR THE
"I AM" WITHIN, AND THE
REAL SELF.

"THE LINE OF INCARNA-
TION", AN IMAGINARY
LINE DRAWN IN CON-
SCIOUSNESS, SEPAR-
ATING SPIRIT AND FORM,
AND THE "i" AND "I AM".

THE ILLUSION OF SEP-
ARATENESS OVERCOME
BY A BALANCE OF FOR-
CES "THE TRUE CROSS.

MAN UPLIFTED AND
CARRYING HIS CROSS.
THE MEANING OF FOUR.
THE TRIUNE SPIRITUAL
MAN MANIFESTING IN
FORM.

THE CROSS ON WHICH
THE SPIRITUAL MAN
IS CRUCIFIED. THE MA-
TERIAL EXPRESSION
EMPHASIZED.

MAN, TRYING TO E-
QUALIZE HIS EXPRESS-
ION, FINALLY DOES—

—BY INTERBLENDING
THE APPARENT OPPO-
SITES, THROUGH A DE-
VELOPED CONSCIOUS-
NESS OF MIND POWER.

9

Introduction

IT HAS BEEN SAID that "Numbers and letters are the signature or name of God," and such indeed is the idea that was entertained of them by many **The Name** ancient peoples, and is still taught **of God** in the Orient, and in this country through the various schools of mysticism.

Numbers and the alphabet are so commonly used by everyone and play so important a part in our lives that it is somewhat **Origin of** curious that we should be so little **Alphabets** concerned about their origin or meaning. Probably most people, if asked where we got our Numbers and alphabet, or why figures are made as they are, or placed in the sequence in which we use them, would say that they did not know and that it made no difference anyway. But neither the names of our Numbers, nor their form and sequence are accidental. They are arranged in a very definite manner which tells a definite, mystical and scientific story. A glance at their history will substantiate this idea, and will serve to indicate their practical influence and their importance to us.

9

Pythagoras, one of the greatest philosophers and Number scientists of ancient Europe, and the founder of our **Pythagoras** present methods of Number interpretation, was born about 580 B. C., as Westcott says, "either at Samos, an island in the Eagean Sea, or at Sidon in Phoenicia." His insatiable thirst for knowledge led him to leave his home as a young man, and he spent many years traveling through the great Eastern centers of learning, India, Persia, and Egypt, absorbing the wealth of wisdom which the sages of those countries had evolved, and finally returning to his native country, Greece, where he introduced the decimal system of Numbers, and the figures which are now commonly called the Arabic numerals, but which he really found as a well-developed system among the Hindoos. From the school he established, the fragmentary references to its teachings which have been filtered down to us through the writings of some of his pupils, and the correlative teachings of the Greeks, Egyptians and Hebrews, the system which the writer has evolved and teaches has been largely compiled, enriched and formulated to a system by his own original researches in connection with Numbers.

In seeking to trace back to the origin of Numbers, we are told by the Hindoos, from whom Pythagoras gained his knowl-
Origin of edge of them, that they were a
Numbers sacred science of the priesthood; and the priests themselves declare Num-

bers to be a direct revelation from the "Devas" or Gods,—symbols of divine realities, disclosing the true nature and plan of the cosmos. This is in striking parallelism to the modern belief among students and teachers of the Ancient Wisdom that Numbers have been revealed to man by the Masters or World Teachers.

The ancient Greek philosophers declared that the decimal system of Numbers was generally believed to have existed for **A Secret** many thousands of years in Egypt, **Science** as a secret and sacred science; and that they formed the esoteric system of which the duo-decimal (or system of twelves) was the exoteric form. This is of special interest in view of the fact that the pyramids are builded upon this system of 12-unit measurements instead of 10. The influence of this duo-decimal system is still an encumbrance of our method of foot and yard measurements, which most other modern nations have long since discarded in favor of the much simpler metric system.

The Greeks and Hebrews both attached profound significance to Numbers, relating them to cosmic forces. They did **Letters and** not use separate systems of letters **Numbers** and Numbers, but their letters were based upon Numbers, and were placed in an order and given a form suggestive of the successive processes of cosmic evolution. The Hebrews particularly emphasized the sacredness of their letters, and gave them an individual significance quite distinct from their combination into words. The form of the letters crudely suggested their meaning, as our present Number system does to some degree. So every Number was a letter also, which relationship gives precedent and reason for the numerical values attributed to the letters of our modern alphabet,—the outgrowth of older systems of letters.

In the Greek and Hebrew names of the Bible there is evidenced a very close relationship between the literal and numerical meaning of the names, and the character or life-expression of the individuals bearing them. Nowhere in this most remarkable Book of numerical symbolism is there a more striking instance of this than that of Jesus, whose name in the original Greek is Iesous (the letter J is not included in the Greek alphabet). The numerical equivalent of the name, easily read by any One familiar with the Greek, since the letters and Numbers are identical, is 888. Among the initiates of the Greek Mysteries or esoteric religion, this Number was always interpreted to mean the "Higher Mind" or "Divine Mind." The Greek word for this idea, "Nous" is the one used in Rev. 13:18 and translated "understanding." It is there contrasted with the "Lower Mind" or "Mortal Mind" which in the Greek is "he Phren," and numbers 666,—referred to in Revelation as the Beast.

Our present alphabet is the outgrowth of these older alphabets and is builded around Numbers. Its formation is not at all

Alphabetical Design a matter of chance, but of design, and the order of its letters follows the order of Numbers. This is the basic idea of word interpretation used at the present time.

Thus, throughout the ages, certain definite ideas have always been associated with numerical quantities, and with the **Symbols of Ideas** figures and letters representing them; and even though everyone may not be prepared to accept the spiritual and scientific accuracy of the attributed **qualitative** values of Numbers, which we have been accustomed to apply only to quantity, there has developed through thousands of years of use, a definite **thought-force** which is universally associated with them. The fact that the same attributes, with very slight variations and modifications, are general among all peoples and races, of all ages, is One of the strongest points in favor of their fundamental verity.

From the foregoing it will easily be seen that Numbers and letters can quite legitimately be said to have an influence upon **Vibratory Influence** our lives, by reason of our own particular association with them through the name we bear and the birthdate we celebrate. That there are other and perhaps weightier reasons for ascribing influences to our names and Numbers goes without saying. But they are of too complex and involved a nature to be extensively dealt with in a book whose circulation is to a considerable extent among beginning students of Number science.

All of us, however, are commonly users or victims of this principle of sound and thought vibration, in a manner which can be **Influence** very easily illustrated. The intona- **of Sound** tion of certain words and phrases believed to have peculiar properties in themselves was quite general among ancient peoples, and survives today in the Orient, particularly among religious cults such as the Buddhists, Brahmans, and Bahais. Catholic priests still intone their prayers in Mass; and even orthodox Protestant clergyman seek to invoke the spirit of humility, reverence, fear or joy, according to their purpose and desire, by reason of the tone of voice they employ. And a secular, and what to the ear at least is a more pleasing demonstration of human susceptibility to sound vibration occurs whenever you listen to a musical composition. We human beings are more susceptible to sound influence than any other mode of vibration with the possible exception of color. Let a skilled musician play a dreamy melody in a minor key and every One of us will respond to the magic of the music by the corresponding mood we express; the feeling of wistful longing, of tender sadness, of memories and of other melodies, and of moonlight nights. This is especially true of the weird harmonies of the Hawaiians. And we have only to hear a stirring march to have our spirits rise and our feet mark time to the martial mood invoked.

You and the Universe, a Book of Numbers

Sound has an actual physiological effect upon us, corresponding to its psychological influence, and will tighten or relax our nerves and muscles according to its mood and tempo.

Of the influence of sound by reason of the thought it invokes, or of sound as a means of conveying thought, a whole book **Sound and** might be written; and each of us, **Thought** from his own experience could write a chapter in it. Why do some of us abhore or adore the name of "Horace" or "Mary" or "Ruth" or "Robert"? Partly because we have probably known someone by those names whose qualities or limitations endeared them to us or aroused our ire; and if not the names suggested, you who read this can well supply others which have a special re-action upon your thoughts. There is a more subtle influence, too, than even personal association, which enters into our attraction to some sounds, and dislike of others, based upon an affinity or lack of it, between the vibrations which are peculiar to each of us as entities, and the vibrations of sound, color, and form; and then there is the more generally understood and accepted influence of sounds by reason of the ideas which we as a people have agreed that they represent. "Love," "hate," "sour," and "sweet" are examples of this.

In many ways, then, there is an actual, tangible, demonstrable relationship between Numbers, letters, and ideas; and all of us **Vibration** act upon, and are being acted upon by the vibrations thus induced.

These vibrations are both mental and physical. All life is vibration, and what we term sound, or color, or Number, are merely different physical octaves of vibration, which are reproduced in the higher octave of thought in the same manner that, when a note is sounded upon the piano, all the corresponding notes in other octaves vibrate in response. So it is that we ourselves are musical instruments, being played upon, and playing with, universal forces in the wonderful Sixty-third octave of thought. (See Table of Rates of Vibration). Our names are the vibratory forces most clearly associated with us from birth to death; they are instilled into the deepest recesses of our minds in childhood, and exert their influence more or less strongly from that time on.

It is an old axiom of the mystics, being corroborated in actually hundreds of ways by modern science, that "as without, so within; as within, so without." There is an absolute correspondence between the outer world of sense, and our personal, inner worlds of body, mind, and spirit. So, by knowing as much as we may of the numerical forces of the cosmos, and their attributes and correspondences, we are also learning something of ourselves, and the forces which manifest through us.

The Cosmos

I. God's Handwriting

NO SYSTEM of words has ever sufficed to produce a Universal Language. As nations and civilizations rise and fall, so does human speech change; and some of the greatest revealments of spiritual truth have been lost with the lan-

The Universal Language guage in which they were written. Nevertheless there is a Universal Language; one which but awaits our understanding to explain all mysteries. It speaks to us through form, sound and color and its alphabet is not One of letters, but One of Numbers. This is not merely a fanciful, poetical idea, but is a scientific fact, which will be further elaborated upon in this and the following lessons. The key to this language is symbolism, which remains unchanged in its universal significance.

Nature makes no mistakes. Her every manifestation presents some divine idea, and unwittingly our actions are made to

Laws and Principle conform to her laws. We imagine we have committed some deed by chance only to discover, as Nature's laws are made clear to us, that what seemed accidental was really the effect of an absolute cause. As we look down the long perspective of the years we discover that what seemed haphazard events are really parts in an orderly procession, leading to One another like links in a chain. Our associates, our vocations, our pleasures, our ideals, our environment, our bodies,—even the names we bear,—are seen to have resulted from the action of definite, unchanging laws, and their re-action in our lives, all in accordance with One fundamental principle of life, which is motion, or vibration.

Scientists tell us that there is only One element in life, and have designated it as motion, and its smallest unit the electron. **The Unit of Form** Thus all substance, all energy, is fundamentally the same, differing only in form, or expression. It is known that the various manifestations of this world in which we live are given their texture, form, resonance, color, and other characteristics, not because of any difference in the quality of the "world stuff" composing them, but because of the difference in the rate of vibration and combination of the One element of which all are made,—the electron.

The quality of all substance is equal, differing merely in the degree of refinement which characterizes its expression. **The Animating Essence** The primary substance which functions through the mineral and vegetable worlds is of the same quality as that which functions through animal and human life. As to the force which produces this element of motion, the scientist is silent; and it has been left for the metaphysician to term this animating esence Spirit, and state the principle that all substance and life is animated by Spirit; that in all things the quality of Spirit is unvaried; and that the degree of refinement in its expression is due to the **quantity** of the animating force.

The investigation which has resulted in these statements regarding substance functioning as matter, has been extended into the subtler realm of Thought, and scientists have actually photographed thought (thereby proving that it has form) and have learned that the same general law of vibration which governs denser forms, applies to the mental realm also.

Vibration of Thought

A wonderful realm of possibilities is opened up by this statement, and science is very slowly systematizing and investigating some of these possibilities, —possibilities which, in many instances, have long been recognized as truths by occult and mystic scientists. One of these truths is that words, which are the vestment of thought, have vibration, the influence and nature of which gave them form and sound. The scientists of old recognized this fact in a much clearer degree than do those of the present day; but probably because they believed the public at large was not ready to receive the knowledge wisely, they concealed it, and taught it to small Numbers of pupils, by word of mouth, lest the unscrupulous and thoughtless should mis-use or deride the power such knowledge afforded. Pythagoras was One of the learned men who

Mystic Knowledge

Why Truth Is Guarded

knew the power of words. He originated a system called the Law of Opposites, based upon the fundamental truth that everything has vibration, and hence can be expressed in terms of Numbers. He discovered that in effect all Numbers reduce to digits, and his system of Number interpretation is so generally recognized as being accurate, that it forms the basis of practically all modern methods.

Having Number, or vibration, everything has form also (either visible or invisible to our limited sight) dependent upon **Motion, Form** the motion of the vibratory **Sound and** units (electrons) composing it. **Color** Another fundamental attribute of motion is sound; hence on some plane all things have sound; and when sound vibrations reach a pitch to which the physical ear no longer responds, they manifest to us through the organ of sight as color.

This may be illustrated by means of a revolving wheel or other object. Turning slowly, its form may be readily **The Effect** seen. Increase the speed and a **of Increasing** low growling note is heard, **Vibration** which gradually ascends in pitch to a hum, then to a shrill, piercing shriek, until the ear will no longer register its tone. (It is interesting to note that some persons respond to lower or higher notes than others; also that we can "feel" notes, sometimes, which are too slow in vibration to be heard. The lower notes on a pipe organ are often inaudible to some persons, and produce an uncomfortable, pulsating sensation which is most disturbing). Revolve the object still more rapidly and a rising temperature is perceptible. As the heat increases the object assumes a dull reddish hue, which changes to orange, yellow, green, blue, purple, and thence to a violet shade, after which color is no longer visible to the human eye. But as the revolution increases ultra-violet rays of light, which can be detected by a camera, are produced. If the rate of motion is still increased, decomposition soon commences, and the object is reduced back to its electronic state, to be rearranged into other expressions of form.

From this simple experiment, investigation may be extended until it is demonstrated that all things have the primary **Fundamental** attribute of motion, form, **Attributes** sound, and color; and that the key to an understanding of these attributes is in Numbers,—since Number describes the most fundamental of all attributes, —vibration. Sometimes only One of these attributes manifests to our physical senses, as in music we hear only sound. But persons of exceptional sight (sometimes called clairvoyants, although the term is often applied to those unworthy the title) can actually trace in space the forms resultant from different tones, and their color.

Possibly this brief consideration of the subject of "God's Handwriting" will serve to

The Only Exact Science illustrate to the inquiring mind of the serious student, the fundamental nature of Numbers. It is generally conceded, even by physical scientists, that they form the only exact science in the world,—mathematics. All other forms of symbolism, like all other sciences, find their origin and basis in Numbers. That stupendous, but simple truth, led the great philosopher, Pythagoras, to declare that the world is builded on the power of Numbers. If that is indeed true, then it is to Numbers that we must turn for a solution of the problems of the world.

II.   What's in a Name

THE UNIVERSALLY accepted method of reducing the letters of the alphabet to numerical quantities is very closely allied to the older Greek, Roman and Hebrew methods, and is generally **The Pythagorean** referred to as the "Pytha-**Method** gorean" method, since it is the adaptation of this system to the English alphabet and language.

The numerical designations follow the order of the alphabet, and with very few exceptions only the digitary value of **Letters and** each letter is used. The reasons **Numbers** for this will become apparent to students as their understanding of the process becomes clear.   The letters and corresponding values are indicated in the following Table of Equations:

1	2	3	4	5	6	7	8	9
a	b	c	d	e	f	g	h	i
j	k	l	m	n	o	p	q	r
s	t	u	v	w	x	y	z	

It will be seen that "o", which is the 15th letter of the alphabet, is given the value of 6 in the Table.   The reason for this is that the digits of 15 equal 6 (1 plus 5 equals 6). Experiment will quickly show that this is true of all the letters.

In working out the numerical equation of a name for analysis the exact method given above is used. Consult the Table of Equations, and place below each letter of the name to be analyzed, the Number which corresponds to it. For example:

Charting a Name

T H E O D O R E   R O O S E V E L T
2 8 5 6 4 6 9 5   9 6 6 1 5 4 5 3 2

The next step is to add the digits of each name separately, thus: THEODORE equals the sum of 2, 8, 5, 6, 4, 6, 9, and 5, or 45. 45 has a digitary value of 4 plus 5, or 9.

The Reduction to a Digit

Hence 9 is the digitary value of the name Theodore. ROOSEVELT equals 50 by the same process. The digitary value of 50 is 5 plus 0, or 5. Hence the digitary value of the name Roosevelt is 5. The addition of the digitary value of both names makes a total of 14, which, by reduction, becomes 5. Thus the total digitary value of the name Theodore Roosevelt is 5. By referring to the interpretations of Numbers Five and Nine, students familiar with the life and character of Theodore Roosevelt can readily see their remarkable application.

This method of numerical equation, i. e., of getting the digitary value of the individual names, and their total value, is called the **Expression or Destiny** Number of the name; and will be found to

The Expression or Destiny

be remarkably accurate, when used with the accompanying method of interpretation, as a guide to the general characteristics of the individual using the name; his likes and dislikes, strong and weak points of character, general temperament, most suitable vocation, and physical susceptibilities.

Important Points In interpreting the Expression of a name, which is the most important point in name analysis, the greatest stress is usually placed upon the total digitary value of the name. In the case of Theodore Roosevelt this would be Five. Secondary importance is attached to each individual name, and lastly the digits corresponding to the letters of each name should be considered. In cases where an individual uses Three names, or Two names and an initial, the process is exactly the same. A middle initial is considered in the same manner as a whole name.

Modifications Each of these processes,—total digitary value, individual digitary value, and value of each letter,—modifies the other. Thus a name adding to Nine, and being composed of individual names and letters in which Two's and Four's predominate, would be somewhat differently interpreted from an Expression of Nine in which there is a predominance of Three's and Five's. These considerations will be an aid to the advanced student in making very accurate and detailed analysis; but for the beginner even the single, final digit of the Expression will be very revealing; and as his experience (and with experience, his increasing confidence both in the accuracy of the author's method, and in himself) is added to, he will find it comparatively simple to judge the manner in which Numbers influence each other, and tend to bring into expression those qualities which are similar, and to modify those which are dissimilar. As a simple rule which the beginning student may safely follow, the total digitary value of the name may be relied upon as being expressive of the principal characteristics of the individual using the name.

As in life generally, the "big" man or woman is the One who includes the most, so in name analysis this law is represented and will reveal itself to be correct, that the name which is most inclusive is strongest. To determine the strength of a name, count the Number of times that each digit appears. The well-balanced, inclusive individual and the name which, numerically considered, is well-balanced and inclusive, will be found in association. The individual will be found to be strongest in the qualities represented by the predominating Numbers, and weakest in those which are in the minimum or totally absent. Sometimes the absence of a digit in the individual Numbers of the name is in a measure compensated for by its appearance in the total digitary value of One of the single names or the whole name.

Strongest Expression

The Inclusion

This "Inclusion" may be quickly found by writing the Nine digits in a row and checking beneath each One the Number of times it appears. Again using the name Theodore Roosevelt as an example, we find the following Inclusion:

Digits: 1   2   3   4   5   6   7   8   9
Inclusion: 1   2   1   2   4   4   0   1   2

Referring to the interpretations of the Numbers, and to their Inclusion in the name, note that Seven, the quiet, retrospective, meditative, rest Number does not appear at all. The Number Five, characterized by excitement, eventfulness, variety, change, adventure, extremes, magnetic personality, and versatility, appears Four times, as does Six,—the Number of strong mentality, of a busy, active career, of poise and balance.

Soul and Body

It has been aptly said that the vowels of a language are its soul. They are the carrying, sustaining power of speech. Every trained singer knows and makes use of the fact. The consonants, contrastingly, are the body, the structure, the garment of speech. In a name, the vowels will be found to express the cherished ideals, the sacred (or "secret") hopes of the individual, the spiritual mission he hopes to accomplish through his life. The consonants will be found to express the impression the individual is likely to make upon other people, their appraisal of his nature and capabilities, from outward appearance and impression.

To discover the Soul, or Ideal, of a name, add the vowels only, but in the same manner that the total Expression of the name was found. To discover the Impression, add the consonants only. In name analysis the greatest stress should be placed upon the Expression; and the Ideality, Impression and Inclusion should be secondarily considered.

Ideality and Impression

There are Three circumstances in connection with the reduction of numerical values to a digit, which do not follow the general rule. If the whole name (the sum of all the individual names) adds to a Number whose digits add to Ten (such as 28, 64, 55) consider the interpretation given for the Number Ten as well as that of One. Ten has a meaning of its own, differing in some respects from that of One, which careful students will find worthy of consideration.

Exceptions —"Ten"

If a single name or the whole name of an individual adds to Eleven, or to a Number whose digits add to Eleven (such as 29, 47, 65, etc.), or to a Twenty-two, the interpretation of these Numbers will be found to apply more accurately to the individual than Two (in the case of Eleven) or Four (in the case of Twenty-two). Often the traits of Two or Four will be present also but in a modified degree. The vowels of both Theodore and Roosevelt will be found to add to Twenty-two;

Eleven and Twenty-two

and in notating the Ideality of the name it would be expressed 22-22 (and not added as digits would be ordinarily). Reduction involving an Eleven is found in the name

```
W  O  O  D  R  O  W        W  I  L  S  O  N
5  6  6  4  9  6  5         5  9  3  1  6  5
         41                         29
          5                         11
```

and instead of reducing the Eleven to Two, and adding the Five and Two to make Seven, the Eleven is retained as a separate Number, and the Expression noted as 5-11.

III.   The Path and Plan

THE BIRTHDATE indicates the Path of Life of the individual; the channel through which the forces of the name will be expressed. It denotes the general trend of circumstances, affairs and **Birthdate** environment, just as the name expresses the personality or consciousness.

Name and birthdate appear to be related to each other in rearing the structure of a life much as the building materi- **Construction** al and the plan by which it is used are related in building a house. No Two individuals seem to have brought with them into life the same material with which to build. Even though in the great fundamentals we are very much alike, we differ endlessly in the details of personality, of impulse, likes and dislikes, talents, whims and ideals. The expression of these is modified and determined to a very considerable extent by the conditions of environment and circumstance into which we are born; and the observation of many students of the law of cycles has traced what is apparently quite a definite relationship between the numerical factors in the date of birth and the circumstances of the life.

The author does not consider the birth-date as of so much importance as the name, in

Circumstance and the "I AM" personal analysis, although the fact that it has an influence is too well authenticated merely to deny. However, personality and what is generally termed character are superior to circumstance ; and the power of applied thought is greater than either of these. Both the name and birthdate represent vibratory forces which are very strongly associated with us throughout life. No One appears to be immune to these influences—and doubtless it is part of the plan of life that we should experience their effects. It does rest with us (the part of "us" which is superior to circumstance, personality and even character, the "us" which cannot be fittingly described by any other name than the individuality, and which we refer to as "I") to determine, if we are alert to our powers, what the effect of the vibrations shall be. The great value in knowing as much about these forces as world consciousness has evolved, is that we may use (or work with) such forces as life presents, rather than being unconsciously influenced by them.

It is quite likely, and to be hoped, that the serious study of earnest students of Number

Growth science will greatly add to our understanding of all vibratory forces, and particularly those of the birthdate ; but present knowledge is sufficient to be very helpful, and to trace corres-

You and the Universe, a Book of Numbers

pondences between dates and circumstances which are of much interest and value.

The generally used method of reducing the birthdate is to consider the month as a **Reducing the Birthdate** Number (according to its place in the calendar) and then reducing day, month and year according to the usual method, thus:

October 5, 1879
10   5   25
1   7

The individual Numbers of this date would be 1-5-7 which would equal 13 or 4. The Number of the Birthpath would then be 4, and the characteristics of that Number will be found to describe the general trend of the outward life, or Path of Life, of the individual.

IV.   Number Interpretations

THIS BOOK is unique in its field as a book of Numbers, in that, so far as can be discovered, it is the only book of its kind that combines an inter-

What and Why pretation of what the Numbers mean, with **why** they are so described.

To the practical mind which is largely concerned with results, the most remarkable feature of name and Num-

"It Works!" ber analysis is that it works! that the meanings ascribed to the Numbers, coupled with the method of numerical equation, are actually reve-lators of the characteristics of the individual with whom they are associated. But to the really serious students of Numbers, those who wish to penetrate back of the realm of effects to that of causes, the cosmic meaning of the Numbers will be of profound interest.

It should be constantly borne in mind that the application of the science of Numbers to personal analysis is only

"The Universal One very small phase of the
Interpreter" potency of Numbers, One application of a great uni-versal principle. The science of Numbers is more than a method of character analysis; it

You and the Universe, a Book of Numbers

is a universal interpreter of life! It is both a science and a philosophy; and those weary searchers for the truth who have made the endless round of theories, speculations and hazards in hope of finding the "pearl of great price" will find "You and the Universe" a logical and inspiring answer to all of their questionings.

In the plan of this book, then, under the heading of each Number, will be given, first of all, its manifold cosmic meanings, and then the application of those meanings to personal analysis.

In personal analysis it will be found that there are Three sub-divisions to each Number-type; a constructive, negative **Threefold** and destructive expression. **Influences** The individual whose Number is Five, for instance, will be observed to express some of the characteristics of each of these Three phases. At his best he will express the constructive characteristics of his type; at his worst he will manifest the destructive tendencies; and in times of repression he will predominantly give evidence of the negative elements of his name. No person expresses only the constructive phase—or either of the other phases to the exclusion of the constructive. All of them exist in all of us. It is most true that we are all mixtures, and that as Henry Van Dyke has said:

"There's so much good in the worst of us,
And so much bad in the best of us,
It ill-behooves any of us
To talk about the rest of us."

Interpretation

In analyzing a name, either your own or that of someone else, think of the constructive elements as those qualities to be encouraged and attained, and of the negative and destructive tendencies as those to transmute, avoid and guard against.

A Word About Health

Following each Number interpretation the reader will find a paragraph relating to the health of different Number types. It is always to be remembered in connection with the adverse physical indications associated with the Numbers, that numerical forces do not cause disease, and that they manifest in disease only in cases where the individual expresses negatively or destructively, thereby bringing upon himself, or causing to re-act within himself, their adverse influence. In other words, the individual (not the vibratory forces) determines health or disease. When the inharmony within the individual causes inharmony of physical expression, it will be found to manifest as indicated in the text. This might be called the physical susceptibility of the type considered.

V.   The Number One

THE NUMBER ONE is primarily the Number of God made manifest. This is indicated in the reference given in the statement of Jesus, that "I and the Father are One," and is so used in many Biblical passages. A somewhat different use **The Number of Deity** of this symbol is shown in the Psalms, as for instance the 19th, whose Number, by reduction, becomes One. In keeping with its Number, the Psalm is devoted to the theme, "The glory of God made manifest."

Man can only conceive of that which is outside his realm of experience, by a comparison of it with what is already **Man and God** known to him. A new mechanical invention can only be intelligently described to him in words with which he is familiar, and must be shown to be related in some manner to some other invention, or to be the application of some mechanical law which comes within the range of his knowledge and comprehension. New words themselves can only be defined in terms of words already known. So it

36

You and the Universe, a Book of Numbers

is quite natural that man's conception of God should be governed by the extent of his own capabilities, and should reflect his own tendencies carried to their ultimate. Thus, to the warring nations of Bible times, the tutelary God, Jehovah, was a warrior. To peace-loving people God was idealized as the God of Peace—a magnified development of their own ideals. God, in the sense the word is customarily used, is merely an idealization of what man believes he should, or would like to be. He is thought of as having the characteristics of man, and the same vices and virtues, only in a greatly enhanced degree.

God the Ultimate

The most primitive idea of God represents Him as man exalted, and is symbolized by the letter I, or the figure 1—the perpendicular line. This symbol is the endless circle broken, defined, or limited to finite space; and is evolved from the physical form of man as the only creature that walks upright and erect. It suggests the unity between man and God.

Symbolism of One

An upright, drawn through the center of the Circle of Being, becomes the symbol of the "fall of man," or the descent of Spirit into Form. The upper half of the circle represents Spirit; the lower symbolizes Man, "created in the image and likeness of God" (in spirit) become God's man-i-fest expression by involution in Form, or mat-

The Fall of Man

You and the Universe, a Book of Numbers

ter. A dot placed at the center of the Circle, at the point between the realms of Spirit and Form, and defining the Two, represents the microcosm, man, of which the Circle itself is the macrocosm. This dot portrays not the outer, physical, imperfect man, but the interior, divine Real Self. The perpendicular extending from the top to the center of the Circle, unites with the dot, which is to say, that the One God is joined to the Real Self of man. The lower, outer, external self of man is represented by the short perpendicular extending upward from the base of the Circle, and reaching toward the dot. This short line is symbolized by the letter "i".

In a diagram thus described are expressed the prime purpose of physical existence, and the meaning of the "fall of man" (his involution in Form). The tendency of evolution and the goal of all progress is to effect a union between the lower nature and the higher, or real, nature, thereby to establish a unity with God, following the example of the Christ, in becoming "One with the Father." By this union with God, man truly expresses the "I", or becomes the great "I AM", and overcomes all limitation. He becomes at One with God, and thus "atones" for the belief in separateness, or existence apart from God. Such belief is the great transgression to which the Psalmist refers (Ps. 19:13).

The
At-one-ment

The Nazarene plainly referred to the necessity of the upraising of the "i" to the "I",

**The
"I AM"**
when he made his Twelve statements of the "I AM", as recorded in the Fourth Gospel.
In view of the significance of
the Number Twelve, these statements increase in significance. The expression "I
AM" is One of the several mystical references
to the One God, the Creator, the Father. It
is the expression for deity especially symbolized by the Number One. Various designations have been assigned to this Number, of
which the author prefers "The Priest" as most
expressive. A priest is supposed to be the
highest representative of God in the flesh,
even as the Number One is the symbol of God-
made man-i-fest, or God given expression
through form.

VI. Number One in Personal Analysis

CONSTRUCTIVELY: One is the Number of creative power, of individuality, of self -confidence and assertion. Just as in cosmic significance it is the primary Number from which the others are Self-expression evolved, so in personal analysis it is expressive of similar elements; the instinct for self-preservation and advancement is strong in the Number One type of person; he interprets life in terms of his own interests; he is likely to insist upon having his own way, and is a forceful, executive, inventive, concentrative influence. He does his best work by centering all of his energies upon One definite line of endeavor, both in thought and action. Strong will power, sincerity, exactness and honesty are characteristic of the constructive One. He seeks a unity and harmony of thought with others, and while his natural tendency is to take the lead in any enterprise with which he is associated, he does not intentionally infringe upon the rights or influence of others. Fair play and "gameness" distinguish his work and play.

Negatively: When persons of the One type slip to the negative phase of expression their lives are characterized

Repression by limitation, a lack of self-confidence and depreciation of their own possibilities. They seem unable to progress in a material way, and while they work very hard to achieve success, their negative, rather hopeless mental attitude is a handicap. Often they become the victims of some infirmity, either physical or mental, which makes them dependent upon another.

Destructively: It will be discovered that just as the negative expression of a Number is characterized by a de-

Domination ficiency of the qualities of the constructive phase, so the destructive expression is caused by an excessive or extreme emphasis of the qualities which normally are desirable. Thus a destructive One makes selfishness of self-interest, domination of leadership, egotism of self-respect. He uses his concentrative powers to advance (?) his personal interests, regardless of their effect upon others, and carries the personal interpretation of life to the extreme of self-centeredness. He is likely to be petulant and irritable, resentful of the success of his associates, and adverse to well-meant suggestions. He is the type of individual upon whose lips you will hear the word "I" most frequently.

Vocational Fitness: The Number One type is best suited to vocations which give him considerable executive pow-**Originality** er but not supreme authority, and the opportunity for original, creative expression. He makes a good manager, foreman, soldier, engineer, inventor, organizer. He is a natural pioneer in any field.

Physical Health: Strong bodies almost always characterize the constructive One. The figurative lack of "backbone" **Back and** not infrequently paralleled by **Lungs** in the negative expression is a literal correspondence in the body,—a weak back or some infirmity due to spinal trouble. A depletion of the vital and nervous forces of the body, due to over-exertion or some form of indulgence (manifest in the destructive influence of One) is likely to manifest through the lungs.

Color: The color particularly associated with Number One is flame. The symbol is a torch. Their lesson is that **A Light** the flaming torch may be **or Flame** either a light to lead and guide One's own life and the lives of others, or a means of razing and destroying and laying waste to the beautiful and desirable elements of life.

VII. The Number Two

WITH the "fall of man," or the descent into Form, there has evolved from Unity, the Duality; and wherever Spirit is made man-i-fest through Form, we find the One become Two. Thus in the al-egory of Creation in the second **Man and Woman** chapter of Genesis we find that space was divided into heaven and earth; time was divided into night and day; and man was made dual in expres-sion,—"male and female created he them",—as were all other forms of manifest life. Ever, throughout God's wondrous creation, this law of duality prevails, expressing as male and fe-male, positive and negative, day and night, summer and winter, heat and cold, good and evil, fire and water, spirit and form.

From these various dualities, there has grown the idea that Number Two is the Num-ber of separation, as it is the **Symbolism of Two** first Number after unity,—the first to break the union of the One with the All. Whereas the Number One expressed the first or most primitive idea of God as man exalted, Number Two distinguishes them,—man **and** God. To the idea of a masculine God, there is added the feminine element, and the expression "Father-Mother God" is evolved, and is also referred to as Father God and Mother Nature.

By the descent of man into Form, a line at right angles to the division between Spirit and matter is formed in the **The Sign** Circle of Being, forming the **of the Cross** mystical sign of the Cross. While we are inclined to think of the Cross as being especially a Christian symbol, (and it has a Christian significance which few followers of the Nazarene seem to understand) it is nevertheless One of the most ancient of symbolical characters. It is particularly associated with the symbolism of the Number Two, in which man is involved in form and is required to express dually the same perfection which was his before the "fall". To do this the upright and horizontal members of the cross must be made equal. The dividing line between the upper (spiritual) and lower (material) parts of the Circle of Being must be through the center of the circle so that Spirit and matter balance each other, and are equal. The lack of balance, in man, of these Two elements which seem to contend within his nature is expressed in the manner in which he has formed the Cross he worships. The lower (material) arm is long; the upper (spiritual) arm is short, showing where he places the emphasis in his life. This is the cross which man must carry,—the cross of disturbed balance,—until he learns the law by which a balance is attained, and the burden is relieved by perfect poise and equilibrium. Because man has so long failed to learn the lesson of the Cross, and has insisted in minimizing the spiritual part and increasing the material elements in life, the Number Two, and the Cross as well, have been considered unfortunate, or "unlucky" symbols. In reality they are both fortunate and unfortunate, good and evil: for that is the great lesson of Number Two,—to discriminate, to choose wisely, to use rather than to abuse.

Two is the Number of Form, because it is through involution in matter, or form, that man must work out the dualities, "to know good from evil" (Gen. 3:22). Even as form is temporal, transient, fleeting, as contrasted to Spirit, which is unchangeable, constant and eternal, so the Number Two is associated with negative, changing, undesirable qualities, as opposed or contrasted to the positive energies of the Number One.

The Opposites

Number Two is pre-eminently the number of sex; and the Cross is a literal representation of sex, and is sometimes formed like a "T" (the sacred letter Tau) surmounted by a Circle. The law of sex is everywhere evidenced in material life, from the spermatozoon and ovum in the most primitive cells, to mankind. It masquerades under varying names, as chemical affinity, attraction and repulsion, positive and negative

The Number of Sex

electricity; and this same law is even made use of in art, though often in a way un-related to the idea of reproduction.

Since sex is the most abused of man's faculties or attributes, it is not surprising that Number Two has long been considered a number of ill-omen. In Genesis the fall of man is attributed to the second human creation, woman; and similarly, Number Two has been called the Number of Woman, and is associated in symbolism with the female deities, as Isis, the Virgin Mary, Rhea, and Vishnu.

Woman

Various significations are given the Number Two in sacred literature, but all bearing out the meanings already suggested. For instance our Bible has Two Testaments, presenting Two tables of law; Adam and Eve, the first Two individuals, had Two sons, One good and One evil; Two angels rescued Lot; the disciples were sent out Two by Two; Two buried Jesus; and there were Two witnesses at the Resurrection and Two at the Ascension. When the same dream was dreamed twice over, it was considered prophetic.

Biblical Two's

The "Two-edged-sword" is referred to in the Bible several times, and presents a most interesting form of the symbolism of the Number Two. Careful examination of the word "sword" itself, is helpful in understanding this symbol. Sword may be

The Two-edged Sword

divided thus: S-word, the power of the Word coupled with the significance of the letter S. (The division is, of course, symbolical rather than etymological). This letter is the Twenty-first of the Hebrew alphabet, which consists of Twenty-two characters. It is called Schin, cognate with our English word "sin," and signifies short-coming, incompleteness, due to its position in the alphabet,—next to the last. The true meaning of sin is incompleteness, imperfection, limitation, or ignorance, and the word should not be associated with the idea of wrong-doing or blame. "S", in combination with other words was sometimes used to designate a negative, destructive action of their meaning. "Word", as used in the Bible, means living substance, the Divine Essence of life, the First Cause: "In the beginning was the Word, and the Word was with God, and the Word was God." (John 1:1). The Sword, then, is the negative, or destructive action of this power of the Word; the power of the Lord (Law) used to destroy, to cleave asunder the good from the bad, the lasting from the temporal, the real from the unreal.

One of the mystical words referring to the Great Unknowable, considered sacred, and only to be spoken upon certain **OM** prescribed occasions, was composed of Two letters, One of which (M) is composed of only straight up-and-down lines, representing the masculine element of life, sturdy, strong, supporting, and rather crude and unbeautiful; the other composed all of curves, symbolizing the feminine element of life, graceful, beautiful, but incomplete of itself, and requiring the support of a stronger force, the O. Together they form the unutterable word OM,—the dual expression of the "I AM".

VIII.   Number Two in Personal Analysis

CONSTRUCTIVELY: Two is the Number of diplomacy, tactfulness, arbitration. As One is masculine and assertive, so Two is feminine and reflective. It is governed more by the affections than the intellect, and the emotions enter **Reflective** largely into the life of a Two. The Two individual is affectionate, considerate, a lover of peace and harmony, willing even to give in a point in argument rather than to have a "scene," faithful in friendship, often secretly undergoing inconveniences and hardships for the real or fancied benefit of a friend. The protective instinct is strong in the Two type of individual, and the only time a Two will fight is in behalf of a friend or loved One, or in defense of the downtrodden or helpless. He is humble, delights in serving those he loves, is quick to anticipate another's wants. The Two, unless other Numbers in association are most assertive, seldom attains great prominence in the world, but he is often "the power behind the throne" of some more dominant personality. His life is lived within himself more than in the outer world. He is patient, unassuming, rather quiet, keen of vision as well as of sight, eager to learn from all sources, and desirous of being friends with all people.

Negatively: The negative Two represents the modified activity of the constructive forces. He is likely to be too **Weakness** emotional, too humble, lacking in initiative, weakly acquiescent to suggestions from others, whether desirable or otherwise, too eager to serve, and likely to idolize those of whom he is fond. He often becomes One of the drudges of life, slavishly subservient to more compelling personalities, degrading high ideals of service by manifesting them in servitude.

Destructively: The Two who allows himself to express destructively is the sullen, disgruntled type, who serves, but **The** serves complainingly. He some- **Misfit** times takes a malicious delight in **Two** withholding the help of which he is capable; and contrastingly will often persist in clinging to someone else when his services are a handicap rather than a help. He is the kind of a Two who does not get on well with other people, who fails to succeed in his own pursuits and resents the success of others. He is irritable, likely to become a sort of petty tyrant to the limits of his small sphere of power, and lacks pride in his personal integrity and appearance.

Vocational Fitness: The Two type makes a good secretary, diplomat, accountant, adjus-

ter of claims, lawyer, or solicitor.
The He is a conscientious parent; is of-
Adjuster ten found in social service work,
and is most efficient when working
with someone else. There are many points of
resemblance between the natures of the Two
and Six, particularly in their fitness for social
reform work, and philanthropic or humanitarian endeavors.

Physical Health: A tendency to headaches is a frequent characteristic of the Two
in poor health; and his physical
Headaches ills generally seem to manifest
through the head, in a mild form
as headaches, and more severely as brain
trouble.

Colors: The color and metal, gold, is
particularly related to Two. Gold is a symbol
of value and of wisdom; and the
Gold mission of the Two, as symbolized
by this metal and color, is "to
transmute base metals into gold"; to discover
the good qualities in all things, and to manifest their relationship to human qualities in
his own expression.

IX.   The Number Three

THREE is, in a sense, the Number of all that pertains to physical existence, being the Number of the dimensions through which Spirit expresses,—Time, Space and Form,—the Three limitative factors of

The Number of Manifest Life

mundane expression, from which, in physical life, we cannot escape. So far as can be learned every manifest thing conforms to this triune formula, although the unaided senses are sometimes unable to detect this conformity in each particular. For instance, we think of music as conforming only to the element of time, and its division into beats; yet the varying pitch traces a specific pattern in space, and its vibrations produce physical disturbances in the atmosphere. If Two tuning forks of the same pitch are placed near each other, and One is struck, the other will be found to give forth a like tone, although undisturbed by any visible agency. Electricity, which apparently transcends time in its operations, does so only by reason of our inability to gain any adequate conception of its extreme rapidity of motion, and actually requires a definite period of time to traverse space. And thought, so apparently intangible and immaterial, is known to have certain well-defined influences upon physical matter and can be photographed. Persons of considerable psychic attainment describe the form and color of thought; and declare the forms it assumes to be so like material forms, as sometimes to be easily mistaken for them.

Each of the Three elements of matter, time, space, and form, expresses a trinity, again emphasizing the persistency of Three as the Number of manifest creation. Time expresses as past, present, and future; space exists as finite, indefinite, and infinite; and form manifests in Three dimensions, length, breadth, and thickness. Space includes something of both time and form, existing in our consciousness only when expressing through One or both of the other members of the Trinity.

Time Space, and Form

Three is called the Number of the arts, measuring the dimensions of architecture; the primary colors, red, yellow, and blue, of painting; and the "Three symphonies which compose harmony in music" (Westcott). It is the Number of mentation, suggesting the thought, the thinker, and the thing thought of. Its relation to the external world is continued in its relation to the interior world of man, designating his triune nature as body, mind, and spirit. Pythagoras called Three the Number of physiology, and some authors have gone to considerable length to justify this association.

The Number of Arts

The Number Three is closely associated with religious beliefs, and trinities abound in all forms of religious worship. The **Trinities** Christians reverence their Trinity of Father, Son, and Holy Ghost. The Egyptians paid homage to Osiris, Isis, and Horus, and related these Three Deities respectively to water (especially the Nile, as the immediate source of so many blessings to them), the earth (particularly Egypt) and the warm, moisture-laden air (the result of the rising of the waters over the earth). Brahma, Siva, and Vishnu, "Creator, Changer and Preserver," each has especial devotees among the Hindus, in addition to those who worship all Three members of the Trinity. Plato evolved what he termed the Divine Triad, of Theos (God), Logos (The Word), and Psyche (The Soul). Mythology has contributed many famous Trinities; among them the Fates, Furies, and Graces, who were each Three in Number, making Three times Three.

Another instance of veneration of the Number is shown in the use, by the Oracle of Delphi, of a tripod from which **Tripods** to deliver her utterances. Less distinguished tripods were much used as supports for bowls of incense, fire, and offerings to the Gods, and in Athens there is a street called The Street of Tripods, because it contains so many of these ceremonial appurtenances.

Three is prominent in Biblical lore. Joshua mentions (Joshua 20:7-8) Three cities

Biblical Three's of refuge on each side of the river Jordan "that the slayer who killeth any person unawares and unwittingly may flee hither" for safety; Bezer (fortress), Ramoth (high places) and Golan (circuit) on the East, and Kedesh (sanctuary), Shechem (back), and Kirjath-arba, "which is Hebron" (alliance) on the West. Ezekiel speaks (14:14) of Three men, Noah, Daniel and Job, "each of whom", as One author says, "saw a creation, a destruction, and a restoration. Noah of the whole world, Daniel of the Jewish world, Jerusalem, and Job of his personal world." The first of these, Noah, "who walked with God," represents the One God, "moving upon the face of the waters." He had Three sons, symbolizing the Three-fold manifestation of God in Creation,—"the One become Three."

Again, with the Number Three, we can trace the law of unity, duality, and trinity in the expression of the sacred word for Deity. First, the I AM, then OM, and third AUM. Even in the form of the characters composing the words the working of the Law is seen. First there was the straight line,—the upright, forceful, masculine element of the I; then the curved line, O, expressing the graceful, gentle, yielding feminine element was placed by the side of M, which is all straight lines; and in the third form of the sacred word, the letter U which is added combines the straight and curved lines of the other Two, and is placed between them to blend the sounds formed in uttering them. (In the Three-fold expression of the word A is usually, but not always, substituted for O).

57

You and the Universe, a Book of Numbers

X. Number Three in Personal Analysis

CONSTRUCTIVELY: Persons of the Three type seek to make their lives expressive of high ideals, and they require an artistic environment for their best personal expression. To some extent this dependence upon external conditions for inspiration is a weakness, but it is sometimes made a strength, when as is often true, individuals of the Three type placed in uncongenial, sordid surroundings, bend their efforts to transform their environment from ugliness to beauty. They are fond of music and often have pleasant singing and speaking voices. They have broad mental vision, which bears with it the tendency toward exaggeration and "make-believe." They have abundant energy which is often used in impractical ways, for they are the dreamers, the visionaries of humanity; but when their powers are applied in a practical manner they accomplish a great deal, and often become veritable geniuses in their chosen field. Thomas Edison, Paderewski, Kubelik, Sargent, and Debs are practical idealists in whose names Three is very prominent.

The Artistic Type

Negatively: The desire for self-expression, strong in the constructive influence, is frequently given inverted expression through self-consciousness, embarassment, and self-depreciation. It is manifested in the "complainer" who attributes his lack of success to his own "spirituality" and his inability to come down to the general level of consciousness. His moral standard is flexible, and he is not above gossip and insinuation.

The Complainer

Destructively: The Three is often superficial, scattering his energy fruitlessly and impractically. He is the dilettant, dabbling into many studies, but not proficient in any of them; lacking the capacity for wholesome manual work, (which he strongly dislikes) and unwilling to perfect himself in any other. He makes mistakes through carelessness and lack of concentration, and tries to shift the blame to other shoulders. He changes his opinions with the prevailing sentiment of others, dislikes to say "No," consequently assumes obligations which he cannot fill, and gains a reputation for irresponsibility.

The Dilettant

Vocational Fitness: The Three type nearly always has musical and artistic ability (dependent in degree upon other forces in the name), is an excellent mimic, and is adapted to professions giving opportunity for original, artistic expression. Writers, artists, musicians, ministers, journalists, de-

The Professions

You and the Universe, a Book of Numbers

signers, actors, reformers, and occasionally nurses and physicians (due to an instinct and aptitude in relieving pain) are found among persons of this Number.

Physical Health: Ills due to some form of dissipation, neglect of the body, and throat trouble, are most prevalent among persons of the Three type.

Colors: The color especially akin to Three is rose-red. This is the color of spiritual love, and it bears a message **Rose-red** which most persons of the Three type need; that of living by love and to love, rather than for love; and of developing the impersonal conception of the love principle that "Service is love in action." Rose-red expresses a highly spiritual conception of life, and, negatively, represents the unfortunate type of spiritually-minded individual who can point "the Way" to others, but lacks the courage and will power to take it himself.

XI. The Number Four

FOUR is the Number of all that pertains to the elements: Fire, Air, Water, and Earth; of the seasons, Spring, Summer, Autumn, and Winter; and of the points of the compass, North, South, East, and West.

The Nature Number — It is associated with stress and strife, with hardships and labor; because it represents the working out of God's plan. Through incarnation, or involution into form, the divine image of man as a triune being is reflected upon the material plane, forming a square which must be "worked out" before man can evolve the conscious expression of his unity with the Supreme.

Recalling to your mind the picture of the Circle of Being already referred to, we may think of the divine creation, "in

Forming the Square — the image and likeness of the Father" (i. e., in Spirit) as a triangle drawn in the upper, or spiritual, half of the Circle, wherein all things must first be imaged before they can be reflected in the lower, or material half of the Circle. The point of the first triangle extends upward, symbolizing Spirit, with the horizontal diameter of the circle as its base; but in its reflection the point extends downward, into matter.

Thus a square is formed, the basis, or foundation, upon which human development must be reared. Mans' divine destiny, then, is to express consciously, through material life, the perfection which he already possesses in Spirit.

This attainment, the perfect evolution of all that has been involved in us spiritually, is **The Soul** the sole (soul) purpose of human **Purpose** experience; and it is because most **of Life** of us do not understand the divine plan, or even know, consciously, that such a plan is in existence, that the working out of life's experiences seems so tedious, difficult, and futile. Illuminated by the light of awakened understanding, the pathway back to God seems far less difficult; for the obstacles, which in darkness seemed so large, are seen in the light of clearer understanding, to be much less fearful than we had supposed. Thus our way is made plain; and the One-ness which manifests through Duality and Trinity is now seen to be evolving back again through the square to Unity. The first Four initiations symbolized by the first Four Numbers, complete the first cycle of developmnt, as is indicated by their additive value; and Unity is attained through the reduction of their sum to a unit: 1 plus 2 plus 3 plus 4 equal 10, which reduces to 1 by adding 1 and 0. All truths may be expressed by Numbers, and the equations of truth are mathematically exact.

Bible references to Four and Forty are numerous and they invariably have to do with

**In
Sacred Writ**
the temptations of human life, and record trying experiences. According to the Bible, transgressions must be paid Fourfold. The Forty years in the wilderness (of sense consciousness); the Forty days of the flood; the Forty days and nights in which Moses received the law on the mount; Goliath's Forty days' defiance of the Israelites; Elijah's Forty days' journey to Horeb; and Jesus' Forty days fasting in the wilderness, are all symbols of the passing of the human soul through Four initiations, each One of which is symbolized by the perfect cyclic Number Ten.

Each of the allegories presents a phase of the soul's supremacy over physical limitations,
**The
Symbolism
Applied**
desires and appetites, which are like a giant Goliath in human consciousness, that the man of God, David, must slay, or triumph over. In the story of the slaying of Goliath the death blow was dealt with a stone in a sling or pouch (a mystical reference to the creative force, and its upliftment to the regenerative plane).

The Four rivers in the garden of Eden (Gen. 2:10-14), Pison, Gihon, Hiddikel, and
Four Rivers
Euphrates, are a symbolical reference to Four great life currents or flows of purified blood ("rivers") supplying nourishment and vitality to ("watering") the garden of Eden (the human body).

The Fylfot, or Swastika, which is but a variation of the Cross, is associated in symbolism with the Number Four, and is universally considered an emblem of good fortune. It represents the working out of the square, or the triumph over matter, and when it is rapidly revolved, tends to reproduce the circle. By certain branches of esoteric teaching it is declared to be related to a magic square of Five, forming Twenty-five squares, of which it retains Seventeen—symbolizing the Twelve signs of the zodiac, the Four elements, and the sun. A most striking fact in regard to the swastika is that it has been used in all parts of the world among all peoples, apparently independent of One another.

The Swastika

The sacred word, which we have traced through Three Numbers, is again discovered to express through Four, in many forms. It has been called Amun in Egypt, Theo in Greece, Allh in Arabia, Dieu in France, Dios in Spanish, Adad in Assyria, Gott in Germany, and Deus among the Latin speaking peoples.

The Sacred Word

XII.   Number Four in Personal Analysis

CONSTRUCTIVELY: Practicality, work, and reliability characterize the Number Four. Persons expressing this Number are usually hard workers (though they do not always choose to be) ; are dependable, industrious, cautious in regard to money matters, desirous of saving and investing, but slow to accept unfamiliar propositions.   They are fond of home life, but sometimes make their homes unhappy by reason of uncontrolled temper and jealousy.   They have a mechanical turn of mind, are analytical, thorough, and practical in the extreme.   They are democratic and patriotic; loyal to home, friends and country.   Their work is nearly always of a nature to meet some need common to the masses of humanity.   Their interests center in practical things, and the phases of life which are the very soul of existence to the Three type, awaken comparatively little response in those most influenced by the Number Four.   They are not necessarily intolerant of, or insensible to, the artistic and spiritual expressions of life, but consider these as merely adjuncts to the more material necessities and conveniences of civilization.   They are very decidedly pragma-

Industry

tists; busy, active workers, who justify all endeavor by the degree that it meets a common need, or as they might express it, a need of the "common people."

Negatively: The apathetic Four is the type whose life expresses a monotone. Negative Four's become the plodders who **Apathy** look upon all expenditure of energy as labor, and who work to live rather than for the joy of it. They have only a fair degree of intellectual force, either constructively or destructively, and are inclined to think life holds but little for them,—the type who "bargains with life for a penny" and gets just that.

Destructively: The ambitious Four who has allowed himself to express destructively is the dissenter. He works hard **The** and is discontented in his work. He **Dissenter** is the clockwatcher, the agitator. He believes himself to be "just as good as any man," and rather defies anyone to say that he is not. He is inclined to over-estimate his importance and worth in the commercial world, and resents the fact that others do not share his personal opinion of his capabilities. When given power he does exactly what he objects to in others, uses his power selfishly, is suspicious of his associates and business-superiors; dislikes being controlled, but seeks to control others. He is likely to have a violent and poorly-governed temper, and to be of a jealous disposition.

Vocational Fitness: The Four type, generally speaking, makes a better employee than employer. He is efficient in work **The** which deals with material commodi- **Artisan** ties,—with things that can be weighed, measured and analyzed. He makes a good engineer, metal-worker, mechanic, electrician, chemist, cabinet maker, builder, architect, foreman, manager, merchant, teacher, business man, or soldier. He is generally suited to work involving great detail, use of the hand as well as the brain, and is efficient where thoroughness, reliability, honesty, and careful workmanship are strongly involved.

Physical Health: Physically the Four is often the victim of the destructive influence of his emotions. He is most suscep- **The Blood** tible to diseases resulting from poor circulation of the blood, maladies which manifest in malignant growths, and occupational diseases.

Colors: Blue and green are related to Four, and bear a message of understanding and supply. Blue, the color of the over- **Blue and** arching heavens, is symbolical of **Green** truth, and its message is expressed in the humble little blue flower, "Forget-me-not,"—to cleave to the fullness of truth, and to speak it kindly. Green is the color of hope, of supply and renewal. It brings prosperity, but will not permit hoarding wealth. Its use and influence as a restorative, renewing, energizing influence are seen throughout nature.

XIII. The Number Five

THE Number Five is the Number of Humanity. It stands midway between the lowest digit (One) and the highest (Nine), even as man stands midway between the animal forces below him, and the spiritual forces above. For that reason **The Number of Humanity** Five is called the Number of Temptation, since humanity is swayed first by One impulse and then by the other, oscillating between Two extremes. By those who consider temptation unfortunate, Five is called an unlucky Number. It might more justly be called a fortunate Number, since it shows the aspiring soul exactly where he stands in the scale of development, presenting the good and the bad, the high and the low, the gold and the dross, impartially.

Number Five is also the Number of Experience, because it is experience which tests our development. "Experiences **Of Experience** are the sacred, purifying fires which remove the dross from the soul." (See "The Tale of the Wimpus," by Ernest C. Wilson). Experience has been described as the purpose of life, again emphasizing the relationship referred to in this lesson.

Five is half of the perfect Number, Ten, suggesting the idea so often advanced by mystics, that man must be united to another of his kind to express perfection, — the basis of the "Soul-mate" theory. Five by itself is said to be incomplete, and must be doubled to be perfect. It is the Number of generation, creative force, sex, and reproduction, and these must be uplifted to the regenerative plane (symbolized by the Number Ten) if man would gain his heritage of eternal life, according to the regenerationists.

Generation and Regeneration

Again the Number Ten, which is divisible equally into Two Fives, may be said to express the dual nature of mankind, as man and woman, in whom Two forces (material and spiritual) constantly contend, until their secret is learned, and they are combined in One,—"reconciled by the cross." It used to be quite generally accepted that man and woman represented separately and specifically these Two forces; woman the spiritual and man the material; whereas most of us have now developed a consciousness that both these tendencies are present in each sex to a greater or lesser degree, and that sometimes the old standard is actually reversed so that there exists such an anomaly as very spiritual men and very material women.

Man and Woman

Physically man is a series of Fives. There are Five extremities to his body; head, arms,

Physical Fives and legs. He has Five fingers on each hand, and Five toes to each foot. He is generally credited with having Five senses, which he only partially uses; and in reality has Two more not yet consciously recognized and developed.

The symbol for Number Five is the Star, resting upon Two points, with the One opposite pointing upward. This sym-

Man the Star bolizes the One (Spirit) which became Two (man and woman) by involution into matter. The Star in the position described is also a very fair representation, or glyph, of man, his feet upon earth, his arms outstretched, and his head upturned toward heaven. Even so does man exist in Two worlds, One of matter and One of spirit, partaking somewhat of both, and only complete when he has been able to disclose to consciousness, their meaning; when he has uplifted matter (or, more accurately, his **idea** of it) to a level with spirit. This thought is hidden in the figure Ten,—the One (God made man-i-fest in man) placed side by side with the Circle (the Creator, the All in All). When the Star is inverted, with the Two points above the One, it becomes the symbol of evil, or Black Magic,—the perversion of that which is spiritual, to gain material power or possessions.

The meaning of "the star" which fell from "heaven unto earth" when the "Fifth

The Fifth Angel Angel (the Angel of Humanity) sounded" (Rev. 9:1-6) becomes clear when we know that Five is the Number of man, and the star his symbol. Man descended from the realm of spirit (heaven) into matter (earth) and was given "the key to the bottomless pit" (the knowledge of good and evil). That is, he was given a knowledge of opposites, that he might manifest his divine power to overcome temptation through experience and supplant evil with good. When humanity opened the bottomless pit, all manner of troubles were loosed,— not that they should "hurt any green thing, neither any tree; but only those men which have not the seal of God upon their foreheads." Man alone must suffer for his indulgence in evil, because he alone has knowledge of it. It was in that knowledge that he became "as One of us," i. e. as One of the Seven Elohim, or Sevenfold God. The seal of God which proclaimed exemption from suffering refers to the pineal gland, or "spiritual eye," which is brought into use by spiritual growth, and is the mark of understanding. He who has spiritual sight, knows that evil is only relative, and that therefore the pit is really bottomless, —"there is no foundation to it." This understanding alone can relieve man from physical suffering. Those who have not the seal in the forehead are not to die (or be annihilated), though they may long for death, but must "be tortured Five months" (must work out their salvation through experience, the Number

Five) ; and experience is the lash that "stings like a scorpion."

Other Biblical Fives are numerous, and significant when read with understanding.

Biblical Fives Joshua killed Five kings and hanged them in a cave (subjected his Five physical senses to his will. They are like kings, and will rule us if we allow them.) Jesus had Five wounds (symbolical of the suffering that accompanies "living in the flesh" before the spirit is resurrected and dominates).

Many names for deity are written with Five letters, and in each instance they refer to a God who incarnated in physical form. For instance, Allah, **Words and Letters** Thoth (Hermes), and Jesus. The ancient Egyptians, with their wonderful knowledge of symbolism, had the greatest reverence for the letter E (corresponding to Number Five) and placed it over their temple entrances. The Hebrews likewise honored their letter He, which was fifth in their alphabet.

XIV. Number Five in Personal Analysis

CONSTRUCTIVELY: Five is the Number of versatility, magnetic personality and intensity. Persons expressing through this Number demand change, variety, travel and excitement. Monotony is insufferable to the Five individual. He

Personality does not like to do the same thing repeatedly. His enthusiasms are intense but changeable. He is interested in everything under the sun, and always seeks to know the "Why?" A love of mystery, adventure and the unknown are characteristic of the Five type. He is often found among students of the occult and mystic phases of life. He is interested in applied psychology, business efficiency and kindred subjects. His tendency is to go to extremes, in thought, action, and apparel. He is vivacious, talkative, energetic, original, creative, ready and willing to help others, fond of companionship, and usually strongly attracted to the opposite sex. An aptitude for learning foreign languages, for memorizing, and for remembering Numbers and faces is a blessing with which the Five type is often endowed. Life usually brings him much travel, and the opportunity to meet many people in all walks of life.

Negatively: When a Five individual slips to the negative plane of expression he often becomes enmeshed in a net of **Obstacles** limitations, many of whose causes can be traced to some abuse on his own part, and some of which seem to be results of a law whose effects alone are seen in his present life. This type of individual finds it difficult to get ahead, but can do so by proving his superiority over the obstacles that beset his path.

Destructively: The tendencies which the Five must guard against are summed up in the Three words, extremes, indulgence **Extremes** and abuse. He is swayed by dual influences perhaps more than any other vibration, and he must resolutely keep to the middle ground of continence in order to enjoy life's best. It is his mission in life to demonstrate the supremacy of the spiritual man over the material man; to express his creative power on the planes of art, science and religion; to avoid the depths and shadows of life by ever climbing upward and facing the light; to avoid the temptations of personal excess or its opposite, asceticism, by living for and serving others. The Two extremes of genius and degeneracy are expressed through Five, and the person strongly influenced by this Number will need to live always at his highest, and to avoid tendencies to lower his personal standard, to procrastinate, to gossip, to prevaricate, and to influence others destructively.

Vocational Fitness: Excellent salesmen, advertising managers, journalists, novelists, surgeons, educators, teachers of **Versatility** psychology, entertainers, and humorists are found among the Five types. They are suited to any vocation offering change, variety, travel, and the opportunity for creative expression; and will prove capable in positions where versatility, adaptability, magnetic personality, good memory, cheerfulness and enthusiasm are required.

Physical Health: The Five individual is most susceptible to ills arising from over-indulgence or wrong influence in **Moderation** the appetites and passions. He will need to guard against habit-forming tendencies (when these are of a destructive nature), should avoid stimulants, and cultivate moderation in diet.

Colors: Pink, midway between the flame color of One and the Red of Nine, is the color which bears a message to the Number **Pink** Five individual. Pink is the symbol of generation, of life, wholesome physical well-being. Its message is to modify the forces of flame and red (which manifest destructively in passion, constructively in abundant energy) and to preserve a balance of all life's forces.

XV. The Number Six

THE Number Six is primarily the Number of Mind. In the diagrams we have described as representing spiritual man and his reflection in matter (the triangle in the upper half of the Circle of Being revolved upon its base, or "reflected" in matter) it will be observed that the line common to both triangles—the diameter of the circle — is longer than the other sides. This line which divides spirit from matter (?), is the Line of Incarnation, and suggests the emphatic stress placed upon physical birth in the development of the individual. Conceiving the sides of the triangle to represent body, mind, and spirit, this line represents body, and mankind generally retains the stress upon his physical nature throughout his earthly life. All history testifies to the fact that mankind has centered his attention upon physical things, and on the gratification of his physical senses, to the detriment of his mental and spiritual development.

The Number of Mind

When this inequality is evened up, and all Three elements assume equal proportions, a very interesting and curious **Evening up our Development** thing occurs. To be contained in the Circle, the Two triangles, which have been made equilateral by adding to our mental and spiritual natures, and diverting or transmuting some of our physical energies into these other channels of expression, must be overlapped. By this overlapping process, which raises the base of the lower triangle and lowers the base of the upper One, a Six-pointed star is formed. This process produces a third division of the Circle in which portions of the Two triangles interblend; and is symbolical of the One element or attribute of man which can combine and unify the Two opposing elements existent within us, matter and spirit. That attribute is Mind. Through first realizing a divine image of perfect expression (the equilateral triangle), then manifesting this upon the physical plane, through our bodily health, and then combining these Two by the power of mind we are enabled to understand the wonderful mystery of their relationship. By this means alone can we know the meaning of the Six-pointed star, the Seal of Solomon, and the emblem of the Jewish faith. Who has this understanding, who can interpret the mystery of Solomon's Seal (the Seal of the Soul-of-man, or sun-man) has wisdom.

Six has often been called the Number of
Motherhood, and doubtless derives this significance from its association with
The Number the power of mind, Wisdom,
of which has been called the
Motherhood World-mother. Six forms the
second of Three series of Three
(the Nine digits) designated as the Triple Triangle. In that sense it would be considered as
having a feminine force. Being Three times
Two, it also bears a close relationship to Five,
the sum of its factors. Six is called the Number of marriage (motherhood suggested
again) the extension or ultimate of the Five
idea, Five being the Number of man, sex, and
creation or generation.

Six is associated with the idea of Labor
(as in the travail of motherhood, or child-
birth) because of the Six days
The Number of labor and the Six days of
of Creation (Gen. 1:31; Exodus
Labor 20:9-11; Luke 13:14); and in a
slightly different aspect the
same thing is suggested in Proverbs, when
Solomon counsels: "Go to the ant, thou slug-
gard; consider her ways and be wise." This
passage suggests the Number Six in various
ways; the ant is a Six-legged creature noted
for its industry; and wisdom is also suggested
by the Number Six. Moreover, the admoni-
tion occurs in the 6th chapter and 6th verse of
Proverbs—a fact to which the Curtisses call
attention. Six seems to be peculiarly associat-

ed with Nature's working forces, and with the elements; and is declared by some writers to be the Number of the elementals. The Book of Revelation tells of the disturbance of natural forces which accompanied the opening of the Sixth Seal of the Book of Life (Rev. 6: 12).

The Number of the Beast

The Number 666 referred to by St. John (Rev. 13:18) strikingly portrays the significance we have ascribed to Six. "Here is wisdom" (Six is the Number of wisdom). "Let him that hath understanding count the Number of the beast; for it is the Number of a man; and his Number is Six hundred Threescore and Six." This is a direct reference to numerical symbolism, and clearly indicates that "the beast" is neither the Pope or the Kaiser, nor any single individual, but refers to "mortal mind," or "error mind" as it has long been called by the mystics. Any "man" who believes in the great illusion of matter is "the beast." There are many such; the vast Number of people who cannot conceive of anything being real which is not tangible to the physical senses, and who believe in an opposition or antagonism between spirit and matter. Whereas we know (and have Paul's word for it in the 18th verse of Chapter Four of Second Corinthians) that the things which are apparently real and can be sensed physically, are actually temporal, and pass away. This belief in the reality (by which is meant permanence of form) of material things,

You and the Universe, a Book of Numbers

and the indulgence in sense gratification to which such a belief leads, are the cause of the "plagues" foretold in Revelation.

XVI. The Number Six in Personal Analysis

ONSTRUCTIVELY: Six manifests in poise, adjustment, strong mentality and service. A person who expresses the constructive influence of this Number is reliable, thrifty, industrious, sympathetic and fond of the association of other people. He is a calm, restful personality whose influence in the sick room is beneficial and restorative. The Six is likely to find his life filled with the demands made upon him by other people, and he is frequently called upon to complete tasks which someone else has begun and left unfinished. His Number corresponds to Saturday in the week, and reflects many of the ideas which are popularly associated with the day,— the busy finishing up of the week's work, and preparation for Sunday. The Six type of individual has been aptly called the comforter. He loves harmony and is distressed by harsh, discordant noises. He is usually musical, with an especial aptitude for stringed instruments, and very often has a pleasing voice of medium range, baritone or contralto, of full, rich quality. He is somewhat artistic, and seeks to combine artistic expression with practical service. He is frequently connected with philan-

Adjustment

thropical or humanitarian projects or institutions, or governmental service.

Negatively: The negative influence of Six is that of unwilling service which makes work labor and service servitude. It **Listlessness** manifests in a tendency to brood over real or imaginary troubles, to morbidity, and lonely introspection. Listlessness, indifference, and a disregard for obligations characterize this expression.

Destructively: The destructive influence of Six manifests in officious zeal, in anxiety and disturbance which makes the individual influenced by Six more **Officious Zeal** restless and disturbing than poised and restful. This type of Six is often found in lodges and fraternities (the Six is usually a "joiner") among those who voluntarily assume duties and then fail to fulfill them. Their intentions are usually the best, which makes them a little more trying to deal with. They fail because they assume more than they can do, overestimating their ability and capacity to serve, in their eagerness to be obliging. They need praise and approval for their happiness, and are very much agitated and distressed when the inevitable criticism of their inefficiency comes upon them. They are incessant talkers, agreeable and social by nature, and their friendliness and extravagant offers of cooperation are often times quite disarming.

Vocational Fitness: The Six individuals make good nurses and physicians, are often found to possess marked musical and **Service** artistic talent, are successful heads of institutions, ministers, social service workers, housekeepers, and practical humanitarians. In business they are frequently found to be insurance men, undertakers, solicitors, lawyers, dressmakers, milliners and housekeepers.

Physical Health: Worry is One of the besetting sins of the Six type, and brings in its wake a long train of ills such as ner- **Worry** vousness, heart trouble, and defective eyesight and hearing. Their worry is usually more for others than themselves, and, as is generally true of worry, accomplishes little good in compensation for its bad effects.

Color: Orange is the color most strongly related to Six, and its message is whole-ness and invigoration,—the blending of **Orange** red (life, love and energy) with yellow (wisdom). It should inspire all those who are working through the influence of the Six vibration to temper their affections, sympathies and energies with wisdom; and to remember that the best way to be of real service in the world is to help others bear their burdens, rather than merely to be relieved of them.

XVII. The Number Seven

THE NUMBER SEVEN is symbolical of fulfilment, fullness, completion. In the soul development of the individual, Seven denotes the completion of the cycle of physical labors, the culmination of the struggle between man's good and evil **Material** physical tendencies. When he **Completion** has labored for Six days (symbolizing cycles, or initiations) there comes a blessed period of rest, when the evolving soul pauses and looks back upon his efforts, as One who, climbing a steep mountain slope, stops to rest and trace his progress, before continuing the arduous climb. If we could not sometimes make these pauses, get our breath, and satisfy ourselves that our efforts have not been futile, it is doubtful if we could retain the courage necessary to complete our journey.

The sages of old recognized this fact, and in the Bible we find Seven days in the seventh month set aside as a time for **Retrospection** feasting (Lev. 23:41); and it is **and** commanded to observe the **Introspection** seventh day and the seventh year as periods of rest. Even as these appointed times for rest were observed objectively, so there is, in the hidden, inner world of the soul, an appointed time for meditation and prayer, symbolized by the Number Seven; preparing the Initiate for the labors of Eight, when he shall meet and face the Dweller; and of Nine, when he must harmonize Four and Five, preparatory to their mergence in Ten.

Seven is the Number of the fullness of all things pertaining to the material world. It Numbers the notes of the musical scale, **Some** the colors of the rainbow and the spec-**Sevens** trum, the vowel sounds of speech, the days of the week and of creation, and the wonders of the world. The seventh son of a seventh son is popularly supposed to possess extraordinary psychic powers. We speak of the Seven seas, and every seventh wave of the sea has an added power. Seven is the common divisor of the time required in the gestation of animals, and children are sometimes born at Seven months.

Many philosophers divide the life of man into Seven parts, comparable to Seven cycles. It is commonly believed by occultists that the Seventy years alloted to man in the Bible are not to be interpreted literally, but symbolize Seven cycles, in each of which man shall pass One initiation, or learn One lesson. The Seven initiations are frequently associated with the Seven vices and virtues, and with the Sevenfold aspects of deity. Shakespeare, in "As You Like It," describes the Seven ages of man, and Jean Ingelow, in "Songs of Seven," similarly described Seven periods in the life of woman.

The human body has Seven obvious, or external parts, enumerated by Westcott as the head, chest, abdomen, Two legs and **Human** Two arms. There are likewise Seven **Sevens** principal internal organs necessary to sustain life. The human head has Seven apertures; eyes, ears, nostrils, and mouth. There are Seven inflections to the voice in articulation; Seven evacuations of the body; Seven great systems of tissue, each of Seven varieties; Seven parts each to the eye, the ear, the heart, the brain, and the nervous system; Seven functions of the body, Seven layers to the skin. While we usually think of having only Five senses, it is pointed out that we actually have Seven, of which the more acute, mental perception and spiritual understanding, remain comparatively undeveloped. The soft tissues of the body, it is estimated, are renewed every Seven months, and every element of the body is completely changed in Seven years.

Madam Blavatsky, in her "Secret Doctrine," and other occultists in other works, tell us that there are Seven great cy-**The Seven** cles in the development of man-**Great** kind upon the earth; each of **Races** which produce a Great Race; and each Great Race, in turn, is subdivided into Seven Sub-races. These Races and Sub-races do not begin and end abruptly, but merge into One another, so that part of humanity belong to One and part to another, according to individual development. The present Great Race, the Ayran, is the Fifth Great Race, and mankind is theoretically in the Sixth sub-division. Most of humanity, however, rather lags behind this position, and is said to be in the Seventh Sub-race of the Fourth, or Atlantean, Great Race.

Theosophists teach that man has Seven bodies; the physical, etheric double, astral, lower mental; higher, buddhic, and the Self, or user of these. Each of these vehicles functions and is focused upon a different plane of matter, having successively higher rates of vibration, and consisting of more attenuated or subtle substance. Ordinarily we are conscious, through the senses, of only One of these,— the physical body; but mediums occasionally see the etheric double and astral forms, and trained psychics have, in some cases been conscious of the higher, less tangible forms. This leads to the statement, also theosophical, that the earth is similarly constituted; that is, that the earth we see, and upon which we function, is but the lowest form of a world chain of Seven planets, each of which is tangible upon a different plane of matter.

The Seven Bodies of Man and Earth

Ancient astrologers considered Seven planets in their calculations, enumerated below, and have formulated correspondences between these planets, (based upon their orbital proportions, and the

Celestial Correspondence

You and the Universe, a Book of Numbers

digitary numerical value deduced therefrom) and the Seven virtues, Seven deadly sins; the vowel sounds, colors, notes of the musical scale, and days of the week.

Planets	Virtues	Vices	Vowels
Mars	Strength	Wrath	o
Sun	Faith	Pride	i
Mercury	Prudence	Idleness	e
Saturn	Temperance	Gluttony	oo
Jupiter	Justice	Envy	u
Venus	Charity (Love)	Luxury	ee
Moon	Hope	Avarice	a

Colors	Notes	Days
Red	do	Tuesday
Orange	re	Sunday
Yellow	mi	Wednesday
Green	fa	Saturday
Blue (Purple)	sol	Thursday
Indigo	la	Friday
Violet	si	Monday

Biblical Sevens are entirely too numerous to be listed in a book of this nature, and their symbolism will be so readily grasped **Biblical** as to make elaboration of the **Sevens** thought involved superfluous in the majority of instances. Beasts to be used for food were taken by Sevens into the ark, others by Twos (Gen. 7:2), giving the Chapter and verse Numbers which locate this reference an especial interest. Gen. 21:23 speaks of Seven ewes given to bind an agreement, and Gen. 29:18 tells of Jacob's Seven years of service for Rachel. The dual aspect of Seven is shown in Joseph's dream of the well and ill-favored kine, and the full and thin ears (Gen. 41) reminding us that Seven is "the fullness of good or evil." In the destruction of Jericho (Jos. 6:4-16) Seven priests bore Seven trumpets Seven days; and after they had encircled the city Seven times the walls fell. When we know that Sheba means Seven, the Queen of Sheba's visit to Solomon, and her endorsement of his learning, assume added significance. She was the Queen of the "south" (I. Kings 10:1 and Math. 12:42) which means the physical or material realm; and her name would signify the perfection of that realm. Thus Solomon is shown, by means of a story, to have passed the exhaustive examination which would require of him the fullness of material wisdom, or perfection in the understanding of material things. Elisha ordered Naaman to be cleansed of leprosy by dipping Seven times in the river Jordan (II. Kings 5:10-4). Jordan is the River of Life; by dipping Seven times into this stream, is symbolized the Seven initiations which everyone must undergo to be rid of mortal limitations, here referred to as leprosy. Solomon's temple (the human body) was finished in the twelfth month of the sixth year, making Seven years. (No doubt this is a thinly veiled reference to the Seven (ty) years alloted to man, for the building of his fleshly temple). See Ezra 6:15. The Book of Esther, with its wonderfully-told story of Ahasuerus, Mordecai and Esther, abounds in Sevens. There were Seven princes, Seven chamberlains, Seven maidens, Seven days of feasting, and it was in the seventh year that Esther came to the king (Esther 1:5-14; 2:9-16). The sixth chapter and 16th verse of Proverbs enumerates Seven things which are "an abomination to the Lord," and in the first verse of the ninth chapter we are told that "Wisdom hath builded her house, she hath hewn her Seven pillars." The Sevenfold Spirit of the Lord is described in Isaiah (11:2-4), as wisdom, understanding, counsel, might, knowledge, fear (respect) of the Lord (Law), and wise judgment.

Revelation is a book of Numbers, and without a knowledge of numerical symbolism **Sevens** cannot be understood. The Seven **of** spirits and Seven stars (3:1) re- **Revelations** fer to the Seven churches which are in Asis (1:4) and these in turn refer to the Sevenfold manifestation of spiritual truth. Paul explains this in the 12th chapter of first Corinthians: "As the body is One and hath many members, and all the members * * * are the One body; so also is Christ (Truth). But all these worketh that One and the selfsame spirit, dividing to every man severally as he will." Other interpretations will suggest themselves to the studen. It would take a book to do them justice.

Andrew Jackson Davis, founder of the Harmonial Philosophy, tells of Seven spheres in the heavenly world, and de-**Other** scribes them. The expression **Religious** "the seventh heaven of delight" is **Sevens** familiar, though only the newer cults seem to have given it serious meaning. The Number Seven has been considered sacred to many of the gods of mythology, and every day of the week has been sacred to some religious system. Wherever One finds mysticism in religion (and where is it not?) there is the Number Seven. The Kabbalah calls it the "great Number of the Divine Mysteries." "The Voice of the Silence" calls the initiate who has passed his seventh degree of initiation "the happy one." The Hermetic philosophy is based upon Seven principles: Mentalism, Correspondence, Vibration, Polarity, Rhythm, Cause and Effect, and Gender.

XVIII. Number Seven in Personal Analysis

CONSTRUCTIVELY: Seven is the Number of quiet refinement, of ideals rather than action, of reflection and meditation. It bears with it the spirit of the Sabbath Day, and suggests the atmosphere of dignity, reserve and reverence in-

Reflection spired by the stained-glass windows of a cathedral. The Seven likes very much to be waited upon and shown attention, to be well-groomed, and to avoid contact with the sordid phases of life. He is often compelled to live a somewhat solitary life due to an attitude of reserve which even his closest friends find difficulty in penetrating. He is talkative at times, and is most so when trying to conceal some emotion which he feels very deeply. The secretive instinct is strong in the Seven type of individual. He is fond of nature, particularly if he may contact it in a way which will not soil his hands, and he loves flowers. He is frequently connected in a business way with enterprises having to do with products of the earth.

Negatively: The Seven type quite frequently makes himself miserable by reason of

The Recluse an exaggerated sensitiveness, and the consciousness of inability to give expression to his friendliness and affection. He feels that, he is mistreated and misunderstood, and is likely to become a recluse, nursing grievances which he would far better overcome and forget.

Destructively: The destructive Seven is very difficult to get along with; erratic, demanding much attention, unwill- **Dogmatic** ing to render service in return, critical and fault-finding; condemnatory of opinions which do not coincide with his own rather narrow views, and aggravatingly supercilious in his attitude toward others.

Vocational Fitness: The Seven type is often a good executive,—an unseen power behind "big business." He is **Conservative** well suited to all kinds of employment which place emphasis upon refinement, artistic ability, and love of beauty. He makes a good florist, horticulturist, manager or proprietor of the so-called art shops, clergyman, organist and writer. He is sometimes found in politics, and in such instances is a factor of the conservative element.

Physical Health: The spleen is the organ of the body most likely to be affected in cases of ill-health, or to cause **Repression** trouble. Contact with nature is needed by persons of the Seven type. Physical difficulties are often the result of mental repression. The natural remedy,

self-expression, is most harmoniously expressed through art, music and writing.

Colors: Steel and purple are related to
Seven. Purple is a symbol of power, and its
message is that power must manifest in service to be constructive.

XIX. The Number Eight

THE Number Eight is the Number of judgment, wherein the human soul is brought before the "power" (symbolized by earthly or material rulers in Paul's Epistle to Romans), and is obliged to face his real self and be judged. This **The Number** "power" is the Dweller on the **of Judgment** Threshold already referred to, and is ofttimes described by mystics as a terrible, awe-inspiring monster armed with a sword (the Two-edged sword of Truth, which cleaves the true from the false) and sitting in judgment upon all who seek to pass beyond Maya, the realm of materiality, into the world of real, or spiritual, things. Paul's letter to the Romans hints at the real nature of the Dweller as "the minister of God to thee for good"; not some exterior, fearsome deity of vindictive power, but the dweller on the threshold of every human soul,—the perfect idea which God meant each of us to express. Sooner or later we become conscious of this divine Self, this perfect pattern by which our life should be drawn; and in the degree that we have or have not been faithful in the expression of this divine Self, we are terrified or unafraid as we face the Dweller.

The symbolism of the mystics accurately portrays, as occuring in the objective or sense world, what actually takes place, **When Self** by analogy, in the spiritual **Meets Self** realm, when man judges himself; when the self meets the Self. No judgment could be more terrible to the erring soul than to see this comparison between what he "might have been,"—his possibilities,—and what he has given expression. So unless he has honestly learned the lessons of the Seven initiations of the material plane, he will be overcome with fear and discouragement; and will once more cast himself back into the material vibrations of the Cycle, or Wheel, of Seven, to review the lessons he should have learned more carefully. Sir Edwin Arnold describes the manner in which this Self-judgment acts, thus:

> "Ye suffer from yourselves. None else compels,
>
> None other holds you that ye live and die,
>
> And whirl upon the wheel, and hug and kiss
>
> Its spokes of agony,
>
> Its tire of tears, its nave of nothingness."

As a part of our divine heritage from the Infinite, we inherently know the Law; and it is this knowledge within us that **The Judge Within** judges, and condemns or justifies. When we know that we have "walked in our integrity" (worked with the Law) we can approach the Dweller without fear, and though we have not entirely "fulfilled the Law," the conviction that we have been faithful to the best light which is ours, gives us courage and fortitude.

Eight is sometimes cited as the Number of Death and Destruction, and in a sense this appellation is correct, for Number Eight means death to the **Death and Destruction** old, the evil, the wrong, the unjust, the undeveloped, to make way for the new, the pure, the good, the just; that the Law may be fulfilled, and Truth glorified. To those who see its influence in the light of the Spirit, Eight signifies justice, opulence, health, renewal, progress. Granted that Eight means death, the thought is not alarming to those who have spiritual understanding, for Death is man's best friend. The writer recalls an experience in which a little creature of the elements came to him, giving the name "Yama," and declaring himself to be a true friend. Learning the meaning of his name (death) was startling; but he still insisted upon the truth of his friendship. This experience was, of course, symbolical: but the writer has often been reminded of its fundamental value. The wise man dies daily,—to old concepts, old habits, old methods of thinking and living, destroying and purging from his life all that he knows to be undesirable, or detrimental to his spiritual progress. He sees that what is death to mortal eyes, is birth to broader vision; and that destruction is, in larger sight, the beginning of a new era of growth.

Eight deals with the hidden, invisible elements of existence; it is potent in its force and power, yet the operation of that **Hidden** potency is known only to the few. **Forces** The sacred literature of the world contains comparatively few references to it; not nearly so many, for instance, as to Seven.

It will be found, however, in both the Old and New Testaments, that very frequently chapters or verses or both, whose **The** digitary value is Eight, deal with the **Psalms** subjects of which Eight is the symbol. The Psalms are replete with numerical symbolism, and their content may be foretold quite accurately by reference to the Number given them. Observe the frequent references to Judgment, Justice, Compensation, the operation of the Law (Lord) and the cleaving of good and evil, in Psalms adding Eight. The first verses of the 8th, 17th, 26th, and 71st; and the 8th verse of the 35th, are all cases in point.

You and the Universe, a Book of Numbers

One of the few direct references to Eight found in the Bible, is regarding the little understood rite of circumcision. It is interesting to observe that the command "He that is Eight years old shall be circumcised" occurs in the 17th chapter of Gensis, 12th verse) ; as by digitary reduction Seventeen becomes Eight. We must refer to other passages to gain an intimation of the true meaning of circumcision; for instance, in writing to the Colossians (2:11) Paul speaks of putting off the sins of the flesh by circumcision in Christ (truth or spiritual understanding) ; and says farther on (3:9) "Lie not to One another, seeing that ye have put off the old man with his deeds; and have put on the new man, which is renewed in knowledge after the image of Him that created him; where there is neither ....circumcision nor uncircumcision, ...but Christ is all." It is plainly evidenced, then, that the real significance of circumcision was not physical, excepting secondarily, but that it was a symbol of "putting off the old man" (the man of material beliefs and tendencies; the undeveloped, limited consciousness) and "putting on the new man, renewed in knowledge after the image of Him" (Christ, Truth,—the symbol of the perfected, spiritual man, of whom we are the intended prototype). When we have put on this new man; that is, when we have entered the enlarged spiritual consciousness of our possibilities and ultimate destiny, "there is neither circumcision nor un-

An Ancient Spiritual Rite

circumcision" for these are of no import.

Eight hundred Eighty-Eight is the Number of Jesus, and forms a contrast with Number Six hundred Sixty-Six, the **Jesus and** Number of the Beast. As previ-**the Beast** ously explained, 666 is the Number of the Lower (or Mortal) Mind, and by the same system of kaballism, 888 is the Number of the Higher (or Spiritual) Mind, so the contrast still holds, and we are given a hint as to the meaning of these Two prominent figures in sacred writ,—Jesus and the Beast.

An Associated Press item, headed "Number Eight Fateful for Kaiser" has recently been given wide publicity, and **Sinister** may prove interesting to the stu-**Influence** dent of numerical values. To those **of Eight** who fail to recognize the Two-fold influence of every Number, positive and negative, constructive and destructive, it may serve to strengthen the prevalent belief in the sinister influence of Eight:

"Two German emperors died in 1888. Two attempts were made on the kaiser's life in 1878. Frederick William Fourth's mental disease compelled him to make way for a regency in 1858. The year 1848 brought revolution. Frederick the Great suffered his severest defeat in 1758 at Hoch-

99

kirch. The Thirty years' war began in 1618. The Great Elector died in 1688, Elector Johann Sigismund in 1608, and Elector Johann George in 1598. The crowning disaster came in 1918."

XX. Number Eight in Personal Analysis

CONSTRUCTIVELY: Eight is the Number of leadership, justice and judgment, commerce and freedom. Persons of this Number seem to desire almost more than anything else, freedom from economic limitation. They are hard **Leadership** workers, with an unusual ability to bring forth in others the expression of their best. They make splendid executives, organizers, and promoters, and are successful in gaining the good will, confidence and co-operation of others. They are good talkers, convincing in the presentation of their ideas; and impress others with their inclusiveness and keen sense of justice and tolerance. Letter-writing is One of their well-developed faculties as a rule. Their lives are generally more governed by intellect than intuition, although they have both in marked degree. They are usually decided in their opinions, have the tendency and ability to win other people to their way of thinking, are dominant, forceful, executive leaders. Life usually associates them with large crowds, people of prominence, frequently civic enterprise, travel, and organization.

Negatively: We are all mixtures of hopes, fears, strength and weakness, and in our times of wavering and uncertain-

Limitation ty we are most likely to express the negative influence of those forces of vibration which our Numbers express. This is true of all people, and all vibratory forces. In the Eight influence it manifests through fear of want, through actual material limitation, the necessity of fulfilling obligations to poor relations, friends in need, and even to people who seem to have no valid claim upon One's energies. The Eight individual who has allowed himself to be dominated by this negative influence is often seen expressing through financial want, struggling against odds, using his fine forces for humble service when he might be occupying a place of affluence which would enable him to render much greater service both to himself and those who have become dependent upon him. He talks too much.

Destructively: The unscrupulous Eight makes domination of leadership much as does the One, but in a more powerful

Domination way. He manifests the extreme and therefore undesirable action of the constructive expression, acting upon hasty judgment, and sweeping others along with him by the force of his personality. He antagonizes by his too blunt expression of opinion, on the One hand, or uses his powers of persuasion to advance selfish interests on the other. In the lower walks of life the

Eight individual sometimes becomes a "confidence man," a pitiful but unscrupulous parody of his higher, true Self, who merits confidence and a wide following.

Vocational Fitness: The Eight is the natural leader and executive. He is suited to **Executive** commercial enterprises wherein a strong personality, persuasive powers, fluent speech, ability as a letter-writer, and organizing or promoting power are desired. He is often very successful in banking, brokerage, stocks and bonds, insurance, and financial projects generally. He is often a powerful public speaker, and is found among the great orators both in religion and statesmanship. The Eight is, moreover, of an inventive turn of mind, and couples with this ability that of utilizing the inventions of other people. Every Eight does not become a great statesman or financier, nor does every Nine become a great artist or writer,—but the influences which will help him to attain to these are present in either case.

Physical Health: The Eight individual is said to be particularly susceptible to diseases which manifest through growths, such as tumors; and to stomach trouble.

Colors: The colors of canary yellow and tan are especially related by Number to Eight. **Wisdom** Yellow is the color of wisdom, and the Eight individual needs to be mindful of the importance of wisely and justly using all of his faculties; to make the symbolism of yellow as it finds corres-

You and the Universe, a Book of Numbers

pondence in him, expressive of its highest power, wisdom,—the priceless possession,— rather than its lowest form of scheming intellect which manifests through what is popularly called "the yellow streak."

XXI. The Number Nine

IN NUMBER NINE are summed up all the forces of the other digits. It is the most powerful Number of all in the sense that it is the most inclusive. Nine is the Number of The Triple Crown, which is the spiritual symbol awarded the Initiate who **The Triple** has successfully passed The **Crown** Dweller. It is emblematical of attainment on all Three Planes of Being, physical, mental and spiritual. When a student at school has creditably completed his course of study, he is awarded a diploma in recognition of his attainments. So likewise, the Initiate in the wonderful school of materal existence is awarded The Triple Crown when he has successfully passed his final examination and has been found worthy.

Nine is a symbol of man having become more than man. The figure Nine is a literal representation of this idea, portray- **Diversity** ing the upright figure of the Initi- **Through** ate, overshadowed by the Circle of **Unity** his Godhood, or merging into One- ness with his Source. The printed form of the figure Nine suggests another phase of its meaning: from the original state of being, i. e., unconscious union with God symbol-

You and the Universe, a Book of Numbers

zed by the Circle, man has evolved his own individuality and has trod the Ninefold path represented by the digits. The passage of the soul from unconscious union with God to conscious union with Him, presents Nine Great Lessons to be learned. The digits are the exterior representations of these initiations, and each One in its formation suggests the lessons which the corresponding initiation requires to be learned.

By many mystics, Number Nine is considered a very difficult Number to work with, and is avoided. The reason is clear **The** since Nine bridges the gap between **Tester** physical and spiritual vibrations. It is the most exacting and the most persistent of all Numbers; it is the supreme Tester, demanding the utmost from the Initiate who would brave its requirements and seek to merge back into the spiritual and perfect At-one-ment symbolized in the Number Ten. He who would add to the Crowns of physical and mental attainment, symbolized in Three and Six, that of spiritual attainment symbolized by Nine, must have come far along the path indeed; he must be firm in his choice between Good and Evil, for Nine is the Number of these Two forces. The great potency of Nine makes it a tremendous power for either good or ill, as the Initiate may choose; and because he sometimes falters and is unable to pass its exacting demands of right-use, yielding to the powers of evil and using for his own rather than the general good, the knowledge he has

You and the Universe, a Book of Numbers

previously acquired, Nine is often called the Number of Black Magic. It has very much the same significance as Five in this connection, but with greater power. Involved in the One are the powers which find their harmonious and balanced expression through Nine. Five stands midway between these Two, and expresses the pivotal point on which the latent possibilities of the One hinge. It is the fulcrum on which the Rod of Power is balanced.

The aspiration of man is revealed in the formation of its figure. The inherent longings and aspirations of man, as he **Aspira-** journeys along the part of attain-**tion** ment, are simply and eloquently portrayed; the Divine urge to become One with God. The fulfilment of that hope is seen when man is raised to the side of the Creator and "walks before Him perfect." This meaning of Nine is accurately presented in Genesis 17:1, "And when Abram was Ninety years old and Nine, the Lord appeared.... and said unto him, I am the Almighty God; walk before me and be thou perfect." The Nine (99—18—9) is commanded to be perfect in Ten; the upliftment of man to Godhood.

Number Nine has been called "bringing to an end," says Westcott, because the human offspring is borne Nine months by **The** its mother, after which it must be **Finished** born again. He might have added **Work** also that it is the last of the digits and culminates the first grand Cycle of Initiation. Frederick L. Rawson, the famous London metaphysician, who has made something of a study of Numbers and their potency, refers very significantly in his book "Life Understood," to the relation of Nine to the 12,000 of each tribe, who were the elect of God. He says "Every group of Numbers can be reduced to its fundamental value by simple addition, namely: adding them together until you have a single digit. For instance, 144,000 adds up to Nine, which refers to the end of the material counterfeit world; this being the special significance of that Number." While we might not altogether agree with Professor Rawson, that the material world is "counterfeit," since it is a manifestation of God's Will, we believe we can see his viewpoint, which is evidently, that the material world is not real in the sense that it is the changeable, evanescent and temporal form assumed by the "One Reality." In that sense we can accept Professor Rawson's interpretation, for by an understanding of the Ninefold path along which we must travel in our "journey back to God," we gradually overcome the many illusions of appearance which the material world presents. By understanding them and realizing that they are evanescent in their nature, we verge upon the great truth that the only reality there is, is God, and that we express that reality through becoming consciously at-one with Him. Because of this conception, the really wise student will not make the mistake of depreciating the wonder and beauty of the material plane, but will appreciate it knowingly; and knowing appreciation is the only true appreciation, after all. Nine represents Unity in Multiplicity; it is the symbol of matter in that it reveals matter in its true nature.

Mathematically Nine is the basis of many interesting combinations. Multiplied by any Number it always gives a product **The Per-** the sum of whose digits will add **sistence** Nine; thus 9 times 5—45—9. If **of Nine** any Number of Two or more figures is reversed and the smaller is subtracted from the larger, the result will be a Number whose digits add Nine; thus 81 minus 18 equals .63, whose digits add to 9. The sum of the Nine digits themselves equals 45, which reduces to 9, and if the Number 123,456,789 be subtracted from 987,654,321, the result is 864,197,532, which result includes all of the Nine digits and adds 45. If the sum of the digits of any Number reduces to a digit other than Nine, that Number will be the remainder from the actual process of division by Nine; thus the sum of the digits of 287—17 —8. If 287 is actually divided by Nine, the quotient will be 31 and the remainder 8. The following interesting computation is variously quoted, and credited to the mathematician, Neuberger:

$$12345679 \times 1 \times 9 - 111111111$$
$$12345679 \times 2 \times 9 - 222222222$$
$$12345679 \times 3 \times 9 - 333333333$$
$$12345679 \times 4 \times 9 - 444444444$$

You and the Universe, a Book of Numbers

```
12345679  x  5  x  9—555555555
12345679  x  6  x  9—666666666
12345679  x  7  x  9—777777777
12345679  x  8  x  9—888888888
12345679  x  9  x  9—999999999
```

XXII.   Number Nine in Personal Analysis

CONSTRUCTIVELY: Nine is expressive of brotherhood, inclusion, generosity, expression through the arts and sciences, public speaking, writing, teaching, and humanitarian endeavor. It is said of the persons whose strongest **The Emotional** Number is Nine that they **Type** attract more love than any other group of people. They have a very strongly emotional nature, and are keenly sensitive to suffering and injustice in the world. They are the friend of the downtrodden, the helpless, the mistreated, whether these express through human form or in the humbler forms of animal life. They are likely to be generous to a fault in big things, and contrastingly mean in little things. They have a keen sense of the comic as well as the tragic phases of life, and are inclined to be swayed from extremes of light-heartedness to deepest melancholy. They are impulsive, and say things they do not mean, often wounding those they love the most. They have a rather unhappy faculty of being oblivious to others' regard for them, and of unintentionally hurting people's feelings. But they are just as quick to try to make amends for their impulsiveness and thoughtlessness, and when their more serious nature is aroused, are capable of much tact and gentleness. Their sensitiveness to the light and shadows of life gives them power of dramatic and literary expression. Nine is the most expressive of all the Numbers, and in interpreting its presence in a name, the analyst must be guided by its combination with other Numbers. An individual whose total Expression Number is Nine, and composed of Four, One, and Four would be quite a different individual from One whose Numbers were Three, Six, and Nine. The former would express more of the democratic, practical, humanitarian ideals of Nine. He would be interested in the economic problem, in justice to the working man, in individualism, reform. Destructively he might be an anarchist; constructively a practical idealist. The Three-Six-Nine individual would be less practical, much more artistic, something of a dreamer, with very high ideals, keen sensitiveness to suffering and desirous of alleviating it, but less practical in his methods. He would probably want to express a spiritual message to the world through the arts.

Negatively: Nine types are not uncommonly cynical, sarcastic, bitter in a consciousness of their own shortcomings **Futility** and those of others; somewhat futilely rebellious toward this condition, but only spasmodically firm in their resolve to express something higher. They adopt the hopeless attitude of never being able to realize their ideals, are often atheists and fatalists, and dispirited in their attitude toward life generally.

Destructively: The destructive Nine is the agitator, the iconoclast, the idol breaker. **The Iconoclast** He often resorts to violent methods to bring about what to him seem the necessary reforms which will save the world from misery and sin, and thus adds to what he hopes to destroy. He is something of a sensualist along with his big ideals, and not above unscrupulous methods of gratifying personal desires. His own world of being is often divided against itself by his contradictory nature; idealistic, inspirational, visionary on the One hand, and despondent, personally careless in the expression of his ideals, temperamental, given to fits of temper and dissipation on the other. He is strongly amative, but his home is often made unhappy by his impracticality, jealousy and a fluctuating personal standard of conduct.

Vocational Fitness: The Nine is the most difficult of all individuals to place. He **The Artist** is changeable and temperamental, usually quite talented, and is capable of expression along most any line which he devotes himself to seriously. He usually finds his greatest happiness—and consequently the best success—in some form of artistic endeavor. He is not always a great artist,—only occasionally,—but he is always artistically inclined, and capable of some expression of this nature. He is particularly suited to humanitarian work, writing, public speaking, the ministry, sculpture, painting, music, invention (due to his strong imaginative ability), teaching or other work bringing him in contract with children, the law, and professional life generally which gives opportunity for the expression of personal ideals. Association with perons who are less inspirational and more practical is a very wholesome modifying influence for persons of these qualities.

Physical Health: The Nine individual is most likely to suffer from heart trouble, faulty circulation of the blood, or **The Heart** from some trouble afecting the genital organs.

Colors: Red and brown are related to the Number Nine, and reflect a great deal of the temperament of Nine. Red is **Red and** particularly the color of life, energy **Brown** and love; the emotional color, related to the blood of the body. It is a color of temper and temperament, of impulse, of brotherhood and universality. As Nine is akin to One, but diverse rather than unified in expression, so red is akin to flame color. It bears the message of flame, but includes more. Brown is the color of both latency and promise, as symbolized by earth and its expression in growth; and with the idea of destruction or decay, as seen in the return of that growth to earth again. Brown very well symbolizes the dual expression of Nine.

XXIII.   The Number Ten

IN NUMBER TEN is fulfilled the aspiration of One. The student will recall that in the Chapter, dealing with the Number One, the symbolism of the "i" and the "I" was explained,—the purpose of life expressed in the endeavor to make

Fulfilment the objective i-nd-i-v-i-dual express the fullness of the I AM.
The "i" suggested this: man reaching upward to a union with the Real Self (symbolized by the dot). This i-dea finds expression through Nine, in which the upright and the dot (here represented as a small circle) are joined,—objective man joined to his Real Self, or attaining the i-deal of physical i-ncarnation. This might well be thought to be the ultimate of material existence, and this achievement, and its representation in Nine, express the i-dea. If man were only man there would be no further accomplishment, but that man is "more than man" is expressed by the Number Ten, for then he is placed by the side of the Great Circle, the All in All, as the manifest expression of all that the Circle signifies. "Then he shall be as One of us." The reason why the Number One is used to express man and manifest God also, is suggested in Number Ten. There is something awe-inspiring in the

realization that this profound mystery, this sublime reality, should be signalled to consciousness in such a simple, commonly-used form as a Number. The student soon comes to realize, if a study of symbolism arouses any response within him at all, that just as he himself is much more potent in possibilities and significance than is usually recognized, so, too, Numbers are far more than mere quantitative marks for counting, but that they are the spiritual symbols of Divine realities. As colors were originally used to convey spiritual ideas, and later became a secular art, so Numbers have "degenerated" (in use) until to most people they convey no other than a quantitative idea.

The Nine digits trace the course of human progress through a Great Cycle of development, suggesting the different **Unity** lessons, experiences and character- **Through** istics of his onward march. The **Multi-** separate steps are completed in **plicity** Nine; they are fulfilled. or united into One great Whole in Ten. This is mathematically proven by the following addition: 1, 2, 3, 4, 5, 6, 7, 8, 9, 10—55—10—1. Thus from unity through multiplicity back to unity the circle of progress is traced; but the Circle is really a spiral which reaches ever higher and higher, so that while circumstances. like the digits, repeat themselves, it is in a different cycle, or upon a different plane.

The Number Ten has many significations, all bearing upon One theme,—Perfec-

Completeness tion through Completion. Ten is the Circle, but more than the Circle. Instead of the original condition of undifferentiated substance which the Circle represented "in the beginning," Ten represents the All in All expressed and unified, made whole,—or "holy." One by the side of the cipher in 10 symbolizes the individualization which has taken place in the circle, through the cyclic development of the digits.

It will be noted, moreover, that Ten is equally divisible into Two Fives. Five is the Number of humanity, of sex. **The Two Fives** In Ten the Two Fives (the Two sexes, man and woman) are joined. This represents the much-abused and degraded idea of soul-mates, or affinities,—the Two become One. Those who ridicule the idea, do so not without reason, for many who grasp at the idea of the "soulmate" theory have little conception of all that it requires of the individual. Before the Two can become One, in the spiritual sense symbolized by the Ten, they must have balanced the Two within themselves as individuals. "The Two must be reconciled in One body by the cross." This is a reference to the mystical process of re-generation, of which Ten is the potent symbol. The sum of the numbers from One to Ten inclusive is 55,—the Two Fives again,— plainly signalling the message first hinted at in the Number Five.

In the pre-Adamic state man is said to

have been hermaphroditic,—"Two in One,"
but he was unconsciously so.
Duality His consciousness was still fo-
and Unity cussed on the spiritual plane of
existence, with little conscious-
ness of his physical realm of being; unconscious of his union in sex just as he was unconscious of his union with God. So as Ten symbolizes the development of a **conscious** union
with God, it also represents a consciousness
of the union of the dual sex principle within
One body. The completion of the cycle
brings man back to the point at which the evolution process started; but with a vast difference in consciousness.

In a study of the Number Ten is revealed
an interesting association with Three, which
plays so prominent a part in all
Ten and the that pertains to life incarnate.
Trinities One of the great religious doc-
trines that is common to prac-
tically all forms of worship is the Trinity. It
is represented in many different guises to be
sure, but in them all the same principle is
manifest. The fullness of life is expressed
when the One has become Two, and the Two,
Three. In religious Trinities this fulfilment
is reached in the birth of the Third member of
the Trinity, the Son; in life, when the Two
are united and produce the Third. But One
step remains to complete the cycle,—the development of a consciousness that the Trinity
of Spirit,—"the man made in the image and
likeness of God," and its reflection in form as

the Trinity of matter, are not separate and opposed to each other, but are in reality One. The conscious union of these Two produces the Square, and completes the Cycle. It may be mathematically demonstrated thus:

1   plus   2   plus
From Unity evolves the Duality, producing
3   plus   4
the Trinity, which forms the Square completing the cycle, and resolving to Unity again.

equals   10   and   reduces   to   1

With the Third Number after One (Four) then, Unity is attained again, showing how fundamental a factor in human **The Law of** life the Trinity is. That this **Trinities** is a repeating law, which may be applied over and over, is indicated by the following table:

1 plus 2 plus 3 plus 4— 10— 1
plus 5 plus 6 plus 7— 28—10—1
plus 8 plus 9 plus 10— 55—10—1
plus 11 plus 12 plus 13— 91—10—1
plus 14 plus 15 plus 16—136—10—1
plus 17 plus 18 plus 19—190—10—1

As Number One is the Number of generation (See Gen. 1:2), and the beginning of physical existence, so Number Ten **Re-** is the Number of Re-generation, **generation** and the fulfilment of physical existence by its conscious unity with Spirit,—the 1 (the manifest) and the 0 (the un-manifest) in harmonious accord.

Much is being written, and much more is being said by advanced thought teachers about re-generation, and the term is usually applied in a very physical sense, as meaning the diversion of the vital force of the body from expression through generation or creation upon the physical plane, and its use in furthering spiritual development. Sometimes this great force is mis-named "sex-force," though of course, the prefix is not descriptive of the force but of the manner of its use. There is only One real force in the universe, whether it manifests through nature in sustaining our human bodies, in mental effort, or in any of the activites and pursuits of life.

Sex Again

Regeneration means exactly what the word indicates; and has no association with human bodies or any other specific thing excepting as we so apply it. In the impersonal sense, it may well be used to describe the wonderful process by which, through physical incarnation, and experience in meeting the problems of life, humanity eventually learns the sublime truth that life is not generated in sin,—no, not even human or so-called "animal" life,—but that all life is essentially pure and Divine, that it emanates from a Source common to all; that it can never, for an instant, have existence apart from that Source, that it cannot be lost, or destroyed, that its purpose is good, and that

Impersonal Regeneration

all of its forms and manifestations whatsoever, are for the purpose of establishing in consciousness, the perfection which already exists in Spirit.

If physical generation be a law of God,—which it most manifestly is,—then it must be good; if it has any existence at all it **Right-** must be good, for to exist, it must **Use** be of God. If it manifests as evil to any individual, the evil is not of the thing itself, but in our attitude toward it, or our mis-use or mis-application of it. "All things are redeemed through right use." Evidently the sin of man has not been in generation upon the physical plane, but his failure to realize that this is only half of a process which is fulfilled by generation upon a spiritual plane,—re-generation. There is nothing evil or wrong about generation except man's ignorance concerning it; his failure to take account of that Power or Force which is back of generation. When he becomes thus conscious; when he fulfills the spiritual demands of his nature as well as those which are material; when he becomes as conscious of the spiritual significance of physical action, as he is of the apparent physical re-action or result,—then he is re-generated; then he will have done that wondrous thing which the Number Ten signifies, consciously, harmoniously associating that which is manifest with that which is unmanifest,—form and Spirit.

XXIV. Number Ten In Personal Analysis

CONSTRUCTIVELY: Ten is closely related to One, and persons in whose name the Number Ten is prominent, will be found to have the general characteristics of One; but they will have in addition qualities not possessed to so great a degree by the persons of the One type. As described in the preceding Chapter, Ten is related to the idea of a consciousness of the spiritual basis of life, and persons whose names come under the designation of the Number Ten have as their mission in life the discovery of this truth. They should seek the good in all things; and demonstrate unity with all people through understanding. The Bahai revelation, founded upon the teachings of a Persian seer, Baha O'llah, particularly expresses the spirit of Number Ten; humility, forbearance, universality, inclusion. Persons of the Ten type will often discover that their lives are largely made up of having to do little things, menial tasks which seem beneath them. They are required to associate with people and conditions beneath them. In this fact is signalled to them their spiritual mission in life.

Principle

Negatively: The Ten who has not awakened to self-consciousness is likely to be

The Ascetic something of a "whiner," making a great deal of the fact that he is imposed upon and that he is not appreciated. He feels that he is superior to other people with whom he is brought in contact, and is likely to consider things of the flesh as being evil rather than accepting them as manifestations of spirit, and therefore good. He is more likely, if religiously inclined, to glorify God by mortifying the flesh than by doing the works of the Man of Galilee.

"Black Magic" Destructively: The Ten who uses his knowledge of spiritual things destructively is sometimes called a "Black Magician" (and persons of this type are sometimes found to express through the Numbers Eleven and Twenty-two, also), using such knowledge of spiritual things as he possesses to wreak personal vengeance upon others, or to gain wealth. Ultimately, of course, such an attitude is certain to have a disastrous reaction, leading to material ruin, mental breakdown, what is called "obsession," insanity and destitution.

Reflective Vocational Fitness: The Number Ten is distinct from other Numbers in that it does not influence the conditions of life nearly so much as it demands an understanding of these. The Ten individual is found primarily in the vocations suited to the Number One Type, but also in vocations related to the Numbers associated with Ten in his name. Ten is more reflective than assertive.

Physical Health: The Ten individuals are most susceptible to illnesses affecting the lungs and windpipe or trachea.

Colors: Flame and lilac are related to the Number Ten; flame color relating Ten to One, and lilac suggesting the **Flame and** violet tint which is symbolical of **Lilac** spirituality and which relates numerically to the Number Eleven.

XXV. The Numbers Eleven and Twelve

BY MANY of the writers of olden times, and in fact by the ancients generally, Eleven was considered as most unfortunate and evil in its significance,—a fact in striking opposition to the interpretation given it, especially as regards its associa-

Unlucky Eleven tion with names, by certain of the present-day teachers of Number symbolism. In reality it is likely that both assumptions are warranted, and that the Number Eleven, like all other Numbers (and everything else in the world), has within it both potencies,—good and ill. Numbers may well be said to be neither good nor evil in portent, except in our attitude toward, and use of, them and their attributed powers. The position of Eleven, between the perfect Number Ten and the Number Twelve,—symbol of grace and understanding,—might well have been sufficient to give rise to the belief in its ominous portent,—since it is neither the One nor the other. It seems as if ancient teachers would gladly have left it out of their calculations. They had associated powerful, sequential ideas with the Numbers on either side of it, and could find no necessity for involving it in their calculations.

The story of the Christ and His Twelve disciples, Eleven alone of whom were faithful,

The Christ Story attaches a special significance to the Number, making it fortunate in the sense that it numbered the faithful, but disastrous in that their mission and that of the Master could never be properly fulfilled so long as the Twelfth remains in disgrace. Thus Eleven and Twelve are seen to have a most intimate association, and their connection in regard to the Christ story presents the great problem which confronts every student and initiate who would walk the Christ-Way. It is One, too, which many, like philosophers (some of them) of old, were inclined to "leave out of their calculations."

Judas Iscariot, the erring disciple, symbolizes the Life principle in man; his correspondence in anatomy is the generative center. It was he who betrayed the Christ, just as he betrays the Christ principle in many to this day.

In the literal story of the Christ and His disciples, there is little doubt that when Judas betrayed the Master it was with no **The Dual Theme** evil intent. He was a shrewd business man. He thought the Christ made very poor use of the marvelous gifts which were His, and that if He would only demonstrate His powers, all the people would acclaim and follow Him. Thus, in his shrewdness, he determined to force the Christ to demonstrate His powers, by betraying Him to His enemies. Any unprejudiced reader of the Gospels can scarcely

127

doubt but that Judas was far from wishing his Master harm. He was really trying to do Jesus a "good turn" by proving to the ignorant unbelievers that the Christ was superior to any power they might seek to use against Him. The stipend he received was too ridiculously small to be considered as a sufficient recompense for the betrayal, had Judas for a moment realized what the terrible outcome would be. It was merely his lack of harmony with the methods of the Christ that caused all the trouble,—not anything essentially evil in Judas.

This exoteric story is closely paralleled by the esoteric meaning of the drama. The **Unity In Diversity** Twelve disciples symbolize the Twelve faculties of mind in the individual. These are all inherently good, but are required to work in close harmony with the spiritual principle,— the God-in-man. Thus they were all working at their duties when the Christ commenced His ministry (or when the Divine nature began to manifest), but it became necessary that all should work together. Thus, Two or more men may all be doing a good work, but through ignorance of the others' efforts, their best intent may inadvertently counteract all the good which might have been accomplished. Until man begins consciously to assume control of his faculties he is very likely to be living in a kingdom divided against itself. All powers must be brought into harmonious accord with the Divine principle. There is no faculty or power or attribute with which we are endowed, whether physically, mentally or spiritually, which is not susceptible to and does not require the direction and guidance of the God within. None of these things are intrinsically evil,— there is no intrinsic evil,—but any good thing when not rightly used, becomes what the world calls evil. So with Judas, the wonderful life principle,—in some ways the most wonderful of all of the Twelve, —this rule must be borne in mind. In Christian theology the whole system of salvation depended upon Judas; and so in mystic symbolism, the whole system of "salvation (or at-one-ment with the Divine Plan) is dependent upon the life principle, for our evolution of the involved powers, with which as souls we are endowed, depends upon our physical birth. Th principle of generation is the universal characteristic of the material world at its present stage of development, and is likely to continue so for some time to come. As the theological plan of salvation required that Judas betray the Christ, so does the Divine Plan require that the principle he represents be "redeemed," or that it be brought into harmony with the guidance of the Christ-within.

Eleven then well symbolizes the transition-period of the soul, when it is merging from disregard or ignorance of the Divine guidance into co-operation with **Transition** it. It is a Number of stress, development, revelation and unfolding spiritual powers. It is a Number of many changes, as the soul progresses from one plane of effort and understanding into another.

Twelve, in some degree at least, suggests the fulfilment of Eleven; the redemption of the erring disciple. It has many **Twelve** astrological associations, but all center around the One idea of the passage of the soul through many influences and conditions toward the achievement of understanding based in experience. References in the Bible, and in mythology, are almost legion,—far too numerous to enumerate here. The Twelve signs of the Zodiac have been correlated with the Twelve sons of Jacob, the Twelve disciples, with colors, birds, animals, etc. Many systems of religion have been based upon Twelve Gods, or a Twelve-fold God. The Twelve simple letters of the Hebrew alphabet have been related to the Zodiacal signs as a part of the secret work of the Rosicrucians, declares Westcott. The Twelve labors of Hercules represent the passage of the sun (son) through the Twelve zodiacal signs.

"The Twelve hours marked on a watch face can be used to find the cardinal points, if the time is correct, and the sun is **Twelve** visible. Lay the watch flat and **Hours** point the hour hand to the sun, and then the South will be half way between that hour and the figure Twelve."

XXVI. Number Eleven in Personal Analysis

CONSTRUCTIVELY: Eleven is the number of revelation, individuality, serenity, poise, practical idealism. Persons bearing this Number are intuitive, get strong "hunches," and if they obey these will find it to their advantage. Their **Revelation** lives often take an unconventional course (though not immoral, except by personal choice), and they will find their best successes come in unexpected ways. They may plan ever so carefully, but find the cases where events transpire as they anticipate are the exception. They are said to be "God's Messengers," to have a revelation of loyalty, service, beauty, industry or some other practical ideal to express to humanity. They are teachers, either avowedly, along the lines of education, hygiene, mysticism, psychology, or the arts and sciences, or unavowedly through example in the fulfilment of some simple task glorified by being well done. Ten, Eleven, and Twenty-Two are all Numbers which express primarily the influence of their digitary value; and many Elevens are very uninteresting, and sometimes uninspired persons; but there is the latent urge within them to be more than their digitary Number indicates. (They are seldom successful in the usual commercial activities; and when through lack of faith in the unexpected,—always their best friend,—they try to conform their lives to the standard of a purely objective, reasoning, business life, they are not successful. They become discouraged from a material viewpoint, find it difficult to regain spiritual faith, and become the most unhappy of people, the worriers.

Destructive: The destructive Eleven is a cynic, super-critical in his attitude toward **The Scoffer** others, ridiculing the spiritual phases of life, and of a decidedly material nature; using his knowledge of spiritual things to exact financial aid from others. He is more concerned in teaching spiritual truth than in living it, and slow to credit other teachers with a more commendable attitude.

Vocational Fitness: The Eleven individuals are religious teachers, authors, reformers, educators, journalists, dram-**The Man With** atists, artists and musicians. **A Message** Sometimes they are found in business, and if they are living at their highest make their vocation, whatever its nature, a means of revealing truth to others.

Physical Health: Elevens are most susceptible to nervous disorders.

Colors: White, yellow, violet and black are all numerically related to Eleven; and each has a message for the Elev-**White, Yellow,** en individual. White signi-**Black, Violet** fies spiritual understanding, which to be true to the highest in his name, the Eleven should see; yellow is symbolical of wisdom; violet of spirituality; and black of consecration.

XXVII. Number Twenty-Two in Personal Analysis

THE READER will note that in Chapter XXV, the cosmic meanings of Numbers Eleven and Twelve are given, and that the interpretations for personal analysis which follow it are for Eleven and Twenty-two. As explained in the earli-

Twelve and Twenty-two er chapters of the book, Eleven and Twenty-two are given special consideration in personal analysis. In cosmic analogy, however, Twelve is of much greater importance than Twenty-two, and is included in this book to complete the presentation of that phase of Number knowledge. In personal analysis Twelve is reduced to Three, following the general rule of reduction. The cosmic meanings given for Twelve have no direct relationship to the meanings of Twenty-two given herewith.

Constructively: Twenty-two is a more practical Number than Eleven in some respects.

Co-operation The spirit of co-operation is prominent, and there is an enthusiasm for the application of abstract truth to the circumstances of life in a very definite way. The constructive Twenty-two individual is scrupulous in his honesty and faithfulness to a trust. He is a natural leader. He has a keen understanding of principle, is apt in illustrating his ideas, particularly in speech; is somewhat given to exaggeration and extravagance, but desires to express moderation, poise and calm. The co-operative urge often leads the Twenty-two to take part in colonization projects, profit-sharing enterprises and kindred undertakings. He is fond of travel, of contacting different races of people, learning from them, and getting their viewpoint. He will·be found to have many things in common with the Four and Eight types.

Negatively: The Twenty-two is not infrequently found in the ordinary walks of life, expressing more of the Four **Despondency** Nature than the Twenty-two, and while still retaining many of his high ideals, and seeking to make them practical, he is weighted down by the feeling that the opposition to his plans is too great to be overcome, and he lacks the courage to face this situation and work it out. Often persons who slip to this negative expression are found with weak bodies, and sometimes with deformities which hamper them.

Destructively: The Twenty-two who permits a lowering of his personal standard of expression, often becomes care-**Self** less in regard to the fulfilment **Satisfaction** of financial obligations. They are full of visionary but impractical schemes, urging and often receiving the co-operation of other people, to the loss of all concerned. Often their intentions are the best possible, and better than they are given credit for. They err in their lack of a sense of responsibility and accuracy, are inclined to overestimate their business ability, and are often deluded by the idea that they are very successful and efficient, but that their failures are due entirely to "luck."

Vocational Fitness: The Twenty-two is suited to work involving the practical application of philosophical and ethical ideas. He has a gift for illustration, which, when the Twenty-two is combined with an artistic Number, frequently manifests in the work of a commercial illustrator, but which more often expresses through the illustration of anecdote or analogy in speech or writing. Twenty-two individuals are good salesmen, promoters, organizers. They are often attracted to partnerships, and work best in association with someone else. They are inventive and creative, and these faculties express along the lines indicated by other Numbers in combination with Twenty-two.

The Illustrator

Physical Health: Illness usually manifests through lung trouble, bronchitis, pneumonia and kindred ills. The inclination to neglect the care of the body is strong, and the Twenty-two should guard against this tendency, taking every opportunity for improving his physical condition.

Neglectfulness

Colors: Cream color and coral are related to the Twenty-two. Cream is the blending

Coral and Cream   of canary yellow and white which symbolizes the blending of wisdom and understanding. Coral is the blending of pink or red and a little of the yellow, and its message is the wise use of the vital energies.

XXVIII.   The Mathematics of Religion

THAT there can be any relationship between mathematics and religion would probably be thought incredible by the majority of people, and yet to any One seeking for such a relationship, there is an abundance of evidence. The recurrence of certain Numbers in Bible narratives; bcomes significant once we admit the possibility of such a relationship, and we begin to wonder if there can be any definite meaning beneath the apparent "happenstance" that there were Twelve disciples of the Nazarene, Twelve Tribes of Israel, Twelve sons of Jacob, and that this Number Twelve is also the Number of definite faculties of mind in man, and of the Twelve chemical elements of the human body. The significant references to Numbers in the Bible are so very numerous, that it would take several books the size of this, to explain all that they refer to. The subject is mentioned here to call the attention of students to the rich field for numerical study which the Bible affords.

Religion and Numbers

The only way we may gain any conception of anything which we have never seen or heard of, is by comparing it with something already known and understood So in attempting to indicate a relation between mathematics and religion, let us begin

Discovery By Analogy

with certain fundamental precepts of each upon which all are informed and from which we may evolve an understanding of their relationship, to the end of greater knowledge and helpfulness, and a more inclusive idea of the unity of all things.

The first tenet of almost all religious beliefs is the existence of an Infinite Being, which for convenience we call God; **The One** that is, we believe in One Great **Power** Force or Power or Principle from which all things are evolved, and of which the phenomena of nature, physical and spiritual, are the manifestation. This manifestation of the Infinite Being is evolved by degrees through infinite forms.

We, as individual entities, are parts of that Power, expressing through a physical form for the purpose of evolving a Divine **Oneness** Idea. All other forms of life, birds, beasts, plants, minerals, likewise express some Divine Idea capable of infinite expansion within itself.

Here the question arises: Shall we then ever completely express the possibilities of that Idea; will we ever become **Perfection** so truly perfect in expression as **and** to lose our identity and merge **Eternity** back ino the Source from which we have sprung? The answer must inevitably be "No"; for just as we are always progressing, so there is an Infinite Idea to express, and infinite time in which to express it. If we were ever completely to express

You and the Universe, a Book of Numbers

perfection our expression would be at an end, the Infinite would be transcended, and there would be an end to infinity, which, of course, would be inconceivable and a contradiction.

With regard to mathematics, let us begin with the simplest forms we know of, and which are familiar to all; that is, **Simple** simple Numbers themselves. The **Numbers** first thing we discover about them is that they are infinite; while they are numbered we cannot count them. They are absolutely accurate, they prove themselves, and yet we cannot say how many there are nor where they stop, and we know inherently that they never do stop. They extend beyond the ability of the mind of men to follow them and reach heights of which we cannot conceive. They are absolutely baffling, and yet they are the only perfect science. They are so simple any One can use them, and yet in the last analysis, they are incomprehensible.

Because they are infinite, let us confine our attention first to the digits, 1, 2, 3, 4, 5, 6, 7, 8, 9. Each expresses some- **The Digits** thing different from the others **and** and yet we may find One thing **Fractions** common to all of them; they all express the Number One in some multiple. Dwelling upon fractions for a moment, such as ½, ⅓, ¼, etc., we find a similar thing is true of them; they all express the Number One in some degree or division.

Now we are ready to commence a rudimental comparison of religion and mathemat-

ics. In religion we have One
Summary Great Power, of which all things
are an expression in some degree;
in Numbers we have One digit, of which all
Numbers are expressions in some degree. In
religion this One Great Power is expressed
through infinitely smaller degrees of itself, in-
dividualized as souls; in Numbers we have
One Number divided into fractions of itself,
which are infinite in degree. The analogy is
perfect. Considering whole Numbers and
fractions together, we may say, in analogy,
that the whole Numbers,—multiples of the
One,—correspond to the objective expressions
of nature, each possessing in a greater or less
degree, the force of the One; and fractions
represent the inherent sparks of Divinity within the individual, capable of infinite unfoldment and expression, but never equaling in
extent the One.

This relationship of fractions to the
Whole, and of individuals (who are just as
truly "parts of one stupendous
God and whole") to God, is so important
Humanity as to be worthy of considerable
elaboration, and while it is inherently simple, we must go a little further into
mathematics than Numbers per se, for an interpretation. You who have studied algebra
are familiar with that illusive mathematical
combination called "The Variable." There
are Two kinds of variables, increasing and
decreasing variables, and perhaps they can be
best understood by illustration. Let us take

the Number One and make of it a decreasing variable. If we start with this digit and regularly decrease its amount according to a definite system, we have formed a **decreasing** variable which approaches as its limit, zero. For instance, divide the Number One by Two, then divide the result by Two, and divide that result again in the same manner. This process may be continued so long as the patience and materials of the student last, and yet there will always be a possibility of continuing the process. The remaining fraction will approach zero by such fine degrees that to all practical purposes it will be the same as zero, yet in reality will never become zero. We may give the illustration objective form by commencing with a piece of paper and cutting it exactly in half; then cut One of these halves in Two equal parts. A continued repetition of the process will produce a decreasing variable which will always leave a tiny remainder if the division is accurately performed.

An increasing variable is just the opposite. Increase a Number by One-half; then add half of the first increase, etc. Us-

An Increasing Variable ing the Number One in this process, we will form an increasing variable which approaches Two as its limit and yet will eternally vary from that exact amount by a fraction.

Humanity is an increasing variable, approaching God as its limit. We human be-

Human Variables ings are differentiated sparks,—fractions, if you will,—of the Great Force we call God. Our development into expression of the Divine Idea each of us represents, must come from within us; must be an increase in expression of what we already possess in involved form. And as in the case of the mathematically increasing variable, the greater the degree of our development, the slower the process. Our soul growth is like cutting a diamond. The largest rough places are cut off first, then gradually the finer angles and facets of the gem are cut and polished until, at last, it expresses as near perfection as human skill can attain. Grinding off the "rough places" of the diamond is comparable to the process by which the "rough spots" in our character are smoothed off. One of the first lessons we learn is not to kill each other. Learning this lesson is a very appreciable advance in our development, and marks a very decided difference in character existing between the man who kills others and the One who does not. Next we learn (not necessarily in the order given) not to steal, and we have added further to our development, although the difference is not so marked. Perhaps next we learn to be truthful and we have progressed still further, but in a less noticeable degree, in exact analogy with the increasing variable which, as it increases in numerical value is increased by decreasing amounts.

The fraction of Deity which we represent

is no doubt infinitesimally small, but for the purpose of our analogy suppose we **Human** assume that we are represented by **Fractions** "½" and then that we have learned a lesson which has increased our expression of the possibilities inherent within us by ½. Our first step in development was then ¼, and this would make our fraction ¾. Assume our next development ½ of the last or ⅛ and this would make our total expression ⅞. Consider our next step in development ½ of the One preceding it, 1/16 and our total development would be 15/16. This process may be repeated in our lives in the same manner and to the same infinite extent, that it can be applied mathematically on paper, and yet our total expression of involved possibilities will never double that of the original fraction. Our ½ will never become One. The tendency is to double such expression, but an infinitesimal fraction always intervenes between the tendency and the ultimate. A wonderful spiritual lesson is to be gained by this analogy in that we may **almost** double our expression, we may **almost** create a counterpart of the Divine Idea we represent,—but not quite. No matter how long the process of development from within is continued, the fraction we represent will never be twice its original value.

Life is so varied in its infinite manifestations, that nature never repeats, never dupli-**Individ-** cates; each form assumed, each idea **uality** expressed, while closely related to every other, is nevertheless, distinctive. We are always separated in expression to some degree, from the Divine Ideas expressed by every other form of life. It is in keeping with this conception of nature's method of individualization, that we, in our progression, however extensive it may be, never overstep the bounds of the individual idea we represent. However great may be our unfoldment, however complete our spiritual attainment, we are always, throughout infinity, distinct in some minute degree, from every other expression of life.

You and the Universe, a Book of Numbers

TABLE OF VIBRATIONS
Whose Effects are recognized and
studied by science
Number of Vibrations
per second.

		Number of Vibrations per second	
1st	Octave	2	
2d	”	4	
3d	”	8	
4th	”	16	
5th	”	32	
6th	”	64	
7th	”	128	Sound
8th	”	256	
9th	”	512	
10th	”	1,024	
15th	”	32,768	
20th	”	1,047,576	Unknown
25th	”	33,554,432	
30th	”	1,073,741,824	Electricity
35th	”	34,359,738,368	
40th	”	1,099,511,627,776	Unknown
45th	”	35,184,372,088,832	
46th	”	70,368,744,177,644	
47th	”	140,737,468,355,328	Heat
48th	”	281,474,979,710,656	
49th	”	562,949,953,421,312	Light
50th	”	1,125,899,906,842,624	Chemical Rays
51st	”	2,251,799,813,685,248	Unknown
57th	”	144,115,188,075,855,872	
58th	”	288,230,376,151,711,744	
59th	”	576,460,752,303,423,488	X Rays
60th	”	1,152,921,504,606,846,976	
61st	”	2,305,843,009,213,693,952	
62d	”	4,611,686,618,427,389,904	Unknown

XXIX.   The Sixty-Third Octave

LIFE PRESENTS an infinitely varied expression of One universal theme, and once we hit upon the basic principle of its expression in any specific manner, we begin to discover analogies in many other phases of expression. Numbers and a **Analogies** few geometrical figures underly all these varied forms, and if we discover their application to any One phase of life we may expect to find that they apply equally well to a great many others.

In mystical literature we read a great deal about "cycles" of development. Ezekiel referred to these as "wheels within **Wheels** wheels." Science parallels this idea **Within** by the law of octaves, which defines **Wheels** an octave as the interval between One and Eight of the scale, or any interval of equal length, and further states that the ratio of a musical tone to its octave is 1:2 as regards the Number of vibrations producing the tones. The Sixty-two octaves of vibration generally accepted by science are given on the preceding page.

It is significant that in this great universe of vibration, whose pulsations are recorded in Numbers of vibrations per **The** second, that the unit of these **Human Heart** rates of motion, reaching from visible movement up through the octaves of sound, electricity, heat, light (color), and the chemical and X-rays, should be the beating of the human heart, which is One a second. Upon this unit of measurement,—itself representing life to us,—the vast systems of worlds in space are builded.

Within these Sixty-two octaves, which represent all science knows of the material universe, there are innumerable correspondences. We may, perhaps, comprehend this most clearly with reference to the octaves which impinge upon our ears as sound. Each octave of sound may be subdivided into Twelve semi-tones, and when any of these is sounded it tends to induce vibration, (and hence sound) in other objects vibrating to the same rate, or to any octave (or multiple) of that rate. If we strike middle C on the piano, all of the corresponding notes in the other octaves will sound a response. We can hear these notes, or responsive vibrations, for about Seven octaves above it. Then the response makes no sensible impression upon our eardrums; although it is possible to shatter a vase (whose tone we may not be able to hear) by "getting its pitch," or sounding its keynote. Following the octaves which we term sound there are a Number whose nature is as yet unknown to science, then follow electricity and heat, and then color; and if the octaves of the tone of C be followed on up through these to the octave which manifests to our sight as light or color, we will find that the

Correspondences

color Red is the Forty-first octave of middle C and that the octave of the spectrum of light is directly related to the Key of C. This is an example of the involved, yet basically simple manner in which all of the octaves are inter-related.

The occult scientists, the metaphysicians, the psychologists, and more slowly the physic-

An Added Octave

al scientists, are gradually coming to discover what they have not yet actually tabulated and added to their "Table of Vibrations," and that is that there is yet another octave of vibration in which correspondences to all of these different octaves are found,—the octave just beyond the realm of what we term phys-ical vibrations,—the octave of thought. Con-sidering the thought realm as the 63rd octave, it is comparatively a simple matter to explain why it is that colors and sounds affect us; and why thoughts affect our physical bodies.

Thoughts are tones in the stupendous 63rd octave, vibrating at the rate of over 9, 000,000,000,000,000,000 vibrations

Thought-Tones

per second. When we sound a clear "tone" in this thought realm, we are awakening correspondent vibrations in every octave of vibration. How important it is, then, that we keep our thoughts attuned to harmony; that we are increasing the harmony in life's octave of ex-pression, rather than the discords! And how easily explainable it is that colors, music, the spoken word, and the thoughts of other people actually do affect us, physically and mentally.

Every time we think a thought we are actually causing a vibration in space, and the force of that vibration in the grand scale of existence is strengthened by the energy we have given it. If our thought is One of healing, everyone who is responsive to that thought will be helped by it. This is the scientific basis of the metaphysical "receptive attitude" in healing. Likewise if we are not well, (which means that we have become responsive to inharmonious vibrations) and we sound the key-note of health, we are strengthening that constructive, harmonious vibration, thereby helping ourselves and everyone who will attune himself to that "pitch" of thoughts.

Thought-Force

F. L. Rawson, in his notable book, "Life Understood" says: "Now we know that a material thought is only apparent vibration," (Mr. Rawson speaks of every thing which assumes material form as being apparent) "that every planet, every star, and every human being has its definite numerical value, in terms of whole small Numbers. Consequently the whole of the material universe is theoretically a system of vibrations, every combination bearing its exact mathematical relationship to all other parts. This is the material representation . . . . of the absolute law, order and system that exists in the spiritual universe which all is governed by God as Principle, and reflects God. Every single thing, therefore, must have its exact and per-

"Life Understood"

fect position and bearing in relationship to all the other spiritual realities, hence the typical significance of each detail.

"The sun and its planets, arranged in the scale of their space relationship to each other, **Universal** exactly reproduce the musical spacing of a fundamental note **Harmonies** and its harmonies. Most probably it will be found before long that the human body, with its heart, represents the sun, and that the arrangement of the electrons exactly repeats the arrangement of the planets. It will be found that everything in the material world is governed by this relationship of the whole small Numbers."

The "eidophone" is a remarkable illustration of this exactness of mathematical relationships. It is so arranged that **The** when sung into, the tones vibrate **Eidophone** a paste-like substance spread over a tightly-stretched parchment, causing forms to appear in the paste, varying with the note or notes sung into it. In this manner various familiar forms, reproducing the outlines of natural objects, such as leaves and flowers, are caused to appear, indicating the vibratory principle which underlies both the sound and its correspondence in nature. When sand is used instead of paste, the figures assume geometrical form. The designs which appear on frosty window panes are often marvelous reproductions of ferns and flowers, landscapes and even animal and human forms; and snow-flakes are one of the beautiful geometrical expressions of the universal vibratory law.

XXX. . The Illusive Future

THE POSSIBILITY of foretelling the future is One that despite innumerable discouragements and disillusions regarding "infallible" methods, man still persists in seeking. It is a great question whether we shall ever be able to know
The what the future holds, except in the
Illusive general sense that we may safely
Future trust whatever comes to be for the ultimate good of mankind, and that it will be the sequential development of what the past and present have developed. However, there is much to be considered in regard to the whole question of past, present and future, which bids fair,—if its deductions are correct,—to alter our conception of these and to bring the future closer to us, in a sense, than it now appears.

Time and space are inseparably related in our conception of them. We measure the One in terms of the other, and in a way they are the same thing, as the following paragraphs may suggest.

Space, as such, has no dimensions. We think of it in terms of Three dimensions, but space itself is infinite. The in-
Space and finite can have no dimensions. It
Dimensions is our consciousness of space which involves dimensions. Claude Bragdon, whose unusual presentation

of the problems of time and space has been notably free from the involved theories and technical phraseology of the academic text-books on the subject, has compared space to the side of a cliff. A cliff does not have to have steps in it to be a cliff. We cut steps into it for our convenience in mounting it. Similarly space of itself has no dimensions, but our consciousness, or conscious use of it, requires these; and we must think of it in terms of dimensions, in order for it to enter our consciousness.

Time is not absolute, but relative. There is no such thing as absolute time. Time is the means by which we measure our perception of things and events. For instance, it takes about Three or Four minutes to play a roll of music on a player-piano in such a manner that we can be conscious of, perceive, or convey from the outer world of Three dimensions to the world within us, the melody which the record presents. We say that the composition is Three or Four minutes long,— which is our measure of the perception of it. If the instrument could be made to play the record in One minute, the relationship of the notes to each other would not be altered. Their relative time-value is determined by the comparative length of the slits in the roll which represent them. The pitch would not be altered, as this is determined by the length and tension of the piano strings. The notes would merely be played in shorter duration

Time and Perception

and more rapid succession, without any distortion of the integral melody itself. The melody is simply an intricate and orderly combination of different proportions, which translated through the mechanism of the piano into a time and pitch, impinges on our ears as melodius sound. It could be translated by a different application of the same fundamental law into a color harmony, or a space harmony, or a combination of both. But when the tempo of the melody as played on the piano is increased to such an extent that the whole roll is played through in, say, a minute, it ceases to be a melody **to our ears,** because our auditory perception is too slow to grasp it. We "get behind," or fail to keep up with the melody, and do not retain—or even grasp—the sound of any single note before the next is played. In order for us to enjoy the melody it must be played at a time-rate which is agreeable to our means of perception. When we experiment with such a means of reproduction as the phonograph, we are confronted with another reminder of our limited perception, for pitch is not a constant, fixed thing on a phonograph, but is dependent upon the turning of the revolving table on which it rests. If the record is played a great deal faster than was intended by its manufacturers, it becomes indistinct, the higher tones become shrill, and even inaudible, so that the beauty of the melody is lost,—not in actuality,—but to our sense of hearing, which is limited to a few octaves of sound.

You and the Universe, a Book of Numbers

Suppose however, that our perception was more extended, and that we could play a Four minute record in One minute, and still retain a pleasurable sense of the melody. Then in One minute we would have experienced the same sense of time as we now perceive in Four minutes. Increase the reproduction to One second, and we would be experiencing in One second what now it takes us Four minutes to enjoy. At that rate our present Seventy years of alloted life would be equivalent to 16,800 years. If we could attain such a development in all our faculties as to permit of instant perception, time and space would be annihilated.

A Faster Time-rate

Dr. G. LeBon says: "What we perceive of the universe are only the impressions produced on our senses. The form we give to things is conditioned by the nature of our intelligence. Time and space are, then, subjective notions imposed by our senses on the representation of things, and this is why Kant considered time and space as forms of sensibility. To a superior intelligence, capable of grasping at the same time the order of succession and that of the coexistence of phenomena, our notions of space and time would have no meaning.... Time is, for man, nothing but a relation between events." (From "The Evolution of Forces.")

Subjective Time and Space

We are limited in our contact with the past and future, and in our full and complete

**Living
in
Eternity** enjoyment of the present, then, by our material appearances of limitation; and as man gradually becomes superior to these limitations, he comes to live less and less in Threefold time,—as past, present and future,—and more and more in the true world of eternality. The great value of limitations is that "in time" they arouse in us a consciousness that they **are** limitations and unreal, thereby helping us to demonstrate our superiority over them.

As we begin to discover the truth of the statement that all life is vibration, and to couple with it the truth that "the **Unity** universe is mental," we are able to perceive the possibility (and hence permit the manifestation of its corresponding reality in our lives) that time **Unity in Time** and space really exist only as a relationship of events, approaching unity in time as we approach it in consciousness, and that the appearance of events and conditions is the diverse manifestation of the One principle.

When we reach that state of consciousness which is One of conscious co-operation with the Divine Plan, our worriment over the future ceases. **Adjustment** Then the various forms of forecasting events will be even less accurate than they seem to be now, for we will become conscious "influences" in the world instead of being so greatly influenced as we are now.

Since life is vibration, and we are for the

most part very susceptible to its influence,—

Aeolian Harps Aeolian harps played upon by whatever winds may blow,—then it is not at all unreasonable to suppose that the vibratory forces of our names influence the events of our life as well as our character, or re-action to events. This idea has no doubt been the motivating force of the many "systems" for forecasting events by means of name analysis with which a public already exhausted and wearied by unsuccessful "methods" of the past has been inflicted. Most of these methods seem to be based purely upon theory, and are of so general a nature in their prognostications, that it is difficult indeed to see that they serve any purpose at all.

Divination Among the ancients, many methods of forecasting events in connection with names and Numbers were used. The common playing cards of the present day are what must seem to kaballists a profanation of a sacred art, for they were long used as a method of divination. A study of their arrangement will reveal many interesting analogies. They have been called the "Gambler's Bible" because so many of their combinations are paralleled in the Bible. The Four suits have been compared to the Four Gospels, and to the Four Hermetic obligations, "To Know, to Dare, to Do, and to Keep Silent," the Thirteen cards of each suit to the Christ and the Twelve disciples, and the Joker to all-powerful God. Various forms

You and the Universe, a Book of Numbers

of permutations, some of them combining the practices of astrology with numerical designations, and all of them purporting to be an aid to knowing the future, to finding lost articles, and to foretelling the winning Numbers in lotteries or winning contestants in a race, are cited in the writings of the Englishman who signs himself "Sepharial." Few, if any, of the methods seem to be of any practical value.

Of modern writers on the subject of events, the author has found the most interest in, and has been most favorably **The** impressed by the researches of **Yi-King-Tao** "Zeolia J. Boyile," author of "The Fundamental Principles of the Yi-King-Tao." In the book named she relates the events of the life to the letters of the individual's name, crediting each letter with an influence for the term of years represented by its Number (using the same Table of Equations as the present writer). More logically than other writers on the same subject, she declares each name to have a distinct influence of its own throughout the life, contrasting with a less developed method which begins with the first letter of the first name and continues to the last letter of the surname, with the cessation of whose influence, presumably, life ceases! With such a method it would surely be wise to have as long names as possible.

The method advanced in the book by Miss Boyile is based upon the idea that a man

Names and Events

named John would be influenced by the conditions related to J in the first year of his life, by O from the end of the first and for the following Six years, by H for the next Eight years, and by N for Five years more. This would account for the first Twenty years of his life. Concurrently with this influence the letters of the individual's other names, middle name if he has any, and mother's surname, would be considered as having an equal importance. And when the influences of a name have been worked through once it is stated that they repeat; that, in substance, the name is a cycle of vibratory force, which is traced as many times as the life of the individual permits.

Perhaps the method will be more clearly understood by the analysis of a complete name. Take, for example, the name John Henry Harlan, with the mother's surname of White.

J	O	H	N		H	E	N	R	Y
1	7	15	20		8	13	18	27	34
21	27	35	40		42	47	52	61	68
41,	etc.				76,	etc.			

W	H	I	T	E	H	A	R	L	A	N
5	13	22	24	29	8	9	18	21	22	27
34	42	51	53	58	35	36	45	48	49	54
63,	etc.				62,	etc.				

The combination of letters influential for any particular year is called the Table of that **Yearly** year. Thus, the Table for the first **Tables** year of John Henry Harlan's life is J-H-W-H, and for the next Three years it would be O-H-W-H, since the J exerted an influence only during the One year. In the fifth year the O would still be in force, the Two H's would be active and the influence of the W would be merging into that of the H following it. The Table of the fifth year would be written

O
H
W to H
H

to indicate this changing of W to H. It would be a period of great personal strain, of uncertainties and fluctuations, all probably concerning money matters, as will be seen by reference to the chart of influences given in the next Chapter.

It must be remembered that Numbers and letters are not actual forces, but are symbols of forces; and that these forces in **Symbols** turn do not indicate fixed, predetermined events in a life, but rather the vibratory conditions which the individual will have to meet in some manner. Since most people disregard such forces and make no attempt to deal with them at all, except as they manifest in events, the distinction between influences and events is not as pronounced to them as it may be made. This explains why the analysis of influences is so very true to the individual's circumstances and experiences. As he becomes master of himself, knowing and using vibratory events, they will come to assume a more modest place in his affairs.

With regard to the accuracy of this method of analyzing influences we have Miss Boyile's statement in her book **Objections** that "a few well-made experiments will absolutely corroborate in every way the accuracy of the statements made" and that she "strongly objects to having the name 'Numerology' connected in any form" with her teachings, toward which objection the author is not unsympathetic, in view of the "weird teachings generally disseminated" under the heading of Numerology, and which he has contacted. The system of analyzing events (which has been adapted from the old records of the Chinese) has proven fascinatingly accurate in the author's own experience, and he presents it here as a phase of Number work which is scientific in its method, checking all results by observation and experience, and is glad to make special reference to Miss Boyile's contribution to an understanding of Numbers.

The purpose of presenting this phase of Number analysis, which to the uninitiated may seem akin to "fortune-tellin," **Knowledge is Power** is to further human understanding of the action of vibratory force. It bears no relation what-

You and the Universe, a Book of Numbers

soever to "fortune-telling;" it does not pretend to foretell the future, but hopes to help students in a knowledge of the forces which they will be dealing with, or have dealt with in the past. If, by the observance of a relationship between letter combinations and occurences and circumstances, students may be guided as to the effects of the forces with which they must consciously or ignorantly deal, such a guidance is no more "fortune-telling" than the anticipation of a storm when the barometer records a low pressure is "fortune-telling."

162

XXXI. Letters and Events

THIS PHASE of name analysis should only be considered when its use can be made constructive. If you insist upon the mistaken idea that the presence of a seemingly undesirable letter in a name is a calamity, and that you are doomed to "bad **Use** luck;" if you permit yourself to be worried by such a situation, leave the analysis of events strictly alone. In the hands of superstitious and fearfulminded people. the analysis of events will do more harm than good; but if you can make it serve you; if you can realize that nothing need be feared when understood, and will let this method of analysis help you, then by all means study it as thoroughly as you can. Work out the Tables for your own and other people's past experiences, and see if you can trace a relationship between events and influences. Be critical; be careful; be accurate. Make notes of your deductions and results; and when you have applied the system to a sufficient Number of cases (always providing you have used the method correctly) you will be in a position to speak from your personal experience concerning its accuracy.

Remember that just as in judging the influences which make for character, each final digit is modified or accentuat-**Modifications** ed by the individual Number of each name, so in judging

162

the influences related to events, each letter of the Table modifies or accentuates the influence of the others.

A produces activity, and bears a relationship to the Number One, representing in charting events much the same influences that One does in personal analysis. It tends to strengthen the activity of all the other letters in a Table.

Activity

B is less forceful than A, inclines to a delicacy of physical health, is noted in relation to a nervous, highstrung condition, and in an undesirable Table of influences,—for instance, One involving the physical health—B is likely to have the same influence as the Number Two. B is sometimes found in marriage tables.

Delicacy

C signifies invigoration, and tends toward a wholesome physical condition. In unfortunate combinations it is said to indicate its presence by throat trouble.

D involves travel or some other movement of a decided, generally pleasant nature, dependent, of course, upon the other letters. When Two are present in a Table it is sometimes an indication of death, or of serious illness.

Movement

E gives eventfulness to all the other letters in a Table, and seems to bring eventfulness by reason of its own force. It is either favorable or otherwise by association.

Eventfulness

F pertains to matter of the heart, either

The Heart   literally or figuratively, has to do with concealment, and bodily health. The author has found it to be present in marriage Tables.

G brings gain, improvement, or recovery.

Gain   With an O it would mean gain in finance, with an R improvement in health; in a marriage Table it strengthens the probability of marriage.

H is related to personal strain, strong emotion or some trouble or situa-

Personal
Feeling   tion in which personal considerations are strongly involved. I seems to involve much the same emotions as H, and is found in Tables of bereavement or marriage, and generally where the personal feelings are brought into prominence.

J brings leadership, willingly or otherwise, almost always advantageously.

K involves travel or change, and is an in-

Change   tensifying influence, hence sometimes manifests through nervousness, though often coupled with invigoration. It is said to bring "success in bold undertakings."

L brings long and short journeys, travel, possibly loss, and self-sacrifice. Two are often found in Tables involving falls or accidents of some other nature.

M produces change; Two are undesirable, and even dangerous. One some-

Two
Dangerous   times merely brings change in the sense of travel, but Two may

indicate serious illness or death,—with an I or H this would suggest bereavement, rather than the personal experience of transition.

N is frequently found in marriage Tables, and also relates to physical health. Two frequently indicate physical inharmony of some kind.

O relates to financial matters, either for good or ill. This is a signification which is quite generally taught by Number **Money** analysts and when O is the first vowel of a name it is considered an indication of success in money matters. O favors adjustment and wisdom. It is either fortunate or otherwise depending upon conditions. With a G it would mean financial gain; with a U, loss.

P brings responsibility and somewhat fleeting power.

Q is generally favorable, tends to recovery in illness, and success in marriage.

R indicates rapidity; it affects the physical body, and its influence reflects in **Rapidity** affairs. Two tend toward illness and accidents—too great rapidity. It is considered usually unfavorable.

S suggests a crisis or climax in any situation which it influences, but for good. It would give emphasis to illness, for instance, but the tendency would be for recovery. It "sharpens, but protects."

T is generally favorable, indicates change of home, travel or marriage.

U almost invariably means loss of some kind or other. The nature of the loss is usually indicated by the remainder of the Table.

V is associated with travel, extravagance and the tendency toward dissipation.

W is an uncertain, wavering influence. Its favorable aspect is travel.

X is said to be generally unfavorable.

Y is indicative of success, favors recovery in illness, and safety on water.

Z is sometimes in marriage Tables, relates to control, and to secret missions.

XXXII. The Fourth Dimension

THE PROBLEM of comprehending the Fourth Dimension is no more difficult, nor essentially different from, the problem presented by an attempt to understand any subject outside our range of conscious experience. We can only **Recondite** fully comprehend what we have **Problems** experienced; yet we have a very fair perception of many things of which we are not directly cognizant, because we have experienced other things to which they are analogous. As an illustration, comparatively few of the earth's inhabitants have ever circumnavigated it; yet they have a fair understanding of the paradox by which we may travel westward to reach an eastern port, and vice versa; and the phenomenon by which a day is gained or lost enroute is made comprehensible to most of us by reason of shorter journeys in which we have had to set the hands of our watch an hour or so forward or back. And even if we have not had that experience, the law and method of the phenomenon have been made clear to us by recourse to our maps and globes, and a little calculation. We accept the evidence on the testimony of reason.

So it is with the Fourth Dimension. It presents no more alarming demands upon our

Absolute Facts credulity than any other idea which we accept without actual personal demonstration of its verity. Much of what is accepted as absolute fact has gained that appelation on less evidence than can be given for the existence and nature of the Fourth Dimension. We speak very glibly of electricity, yet we know only the phenomena of its presence, not itself. A similar method of cognition must be employed to make the acquaintance of the Fourth Dimension.

For so long people have wished to impress others with the profundity of their thought by **A Mis- apprehension** talking in a very wise manner, but actually very foolishly, about the mysterious Fourth Dimension, that many students still hold the opinion that the Fourth Dimension is a very obscure and recondite subject, which is wholly impractical, and of interest only to college professors and others who are supposed to revel in contriving problems they cannot solve. The Fourth Dimension has proved a convenient alibi for much incomplete knowledge, and the speculation it has aroused has wasted much time, paper and ink,—doubly wasted, since not only have the explanations (?) been incomplete, but they have dealt with a subject which was considered even by its devotees, as merely a sharpener of wits, and of no material consequence.

Unless the Fourth Dimension is a vital,

potent, practical factor in human life, it is a waste of time to attempt to solve its mysteries,—if, indeed, as many people doubt, it has any existence at all outside the thoughts of a few hare-brained people. But if it is a reality, if it has an importance in the solution of human problems, then it is very desirable that we know as much about it as possible. It is on this latter theory of values, that this consideration of the subject,—necessarily very brief,—is presented.

A Question of Values

The only way the human mind can comprehend the unknown or unexperienced, is by comparison with the known or experienced. Whether it be a new form of government, or a strange philosophy, or an improved mousetrap, the same rule applies: its unknown mysteries must be connected, in thought at least, with what is already familiar or acceptable to us. Thus we gain our conception of the Fourth Dimension by reviewing our already possessed knowledge of the Three dimensions, and following in thought the logical path which an extension of that knowledge would trace.

The Unknown

Conceive, if you can, a world of only One dimension, length. The nearest approach to this, in the Three-dimensional world through which our consciousness functions at present, is a piece of string of interminable length. Of course it has a

A One Dimensional World

degree of width and thickness, but so comparatively small in proportion that we can mentally, at least, disregard it. Suppose further that in the world we have imagined, intelligent, living, thinking entities exist. Such entities could have no conception of up and down or across, for their existence would be One of linear extent only. Their bodies would have only length, and they would exist side by side in their world like beads on a string, or links in a chain. Each entity would be bounded on both sides (in the only direction of which he would have any consciousness) by other entities. If his neighbors took a notion to move in One of the Two directions possible to them, he would have, perforce, to move with them; he could not go over, around, or under them.

Now let us leave to its own devices for a time, this One-dimensional world, and investigate another world of vastly extended possibilities,—a Two-dimensional world, comparable to an immense plane surface, like One side of a piece of paper. Conceive it, too, to be peopled by intelligent beings. They would have the inestimable advantage over their One-dimensional cousins, of being able to move around each other, and of thereby selecting their associates—with this possibility, however; that if One of these creatures, on mischief bent, could form a circle of itself, it might surround a fellow creature,

A World of Two Dimensions

and make it a prisoner. The confined creature could have no idea of stepping over the boundary formed about it (or him) because a Two-dimensional consciousness could not conceive either of "stepping" or "over." Such expressions would be as meaningless as the "Fourth Dimension" to us. The Two-dimensional entity would "know" by the evidence of reason and experience, that there could be no third dimension, just as One-dimensional entities would "know" that a second dimension was the hallucination of an unbalanced mind (if balance applies to One dimension) and we, wonderfully intellectual creatures that we are, endowed with marvellous faculties, "know" that there cannot be a "Fourth Dimension."

But again let us suppose,—that creatures of Two and Three dimensions could observe and affect the One and Two dimensional worlds, respec-

One and Two Dimensional Miracles tively. The Two dimensional entity could easily cause the One-dimensional creature to appear and disappear from the string world, by merely doubling the "string" upon itself. In the Two-dimensional world a similar phenomenon could be made to occur through the agency of a Three-dimensional being, who could cause appearances and disappearances in different parts of the plane by the simple device of folding it upon itself.

Fortunately for the clarity of the author's text, there exists in our Three-dimensional

A Two Dimensional Oddity

world, One thing of only Two-dimensional proportions. It is exactly like what we have reason to suppose a plane-entity would be like. Such an entity could have no color which would be visible to its associates, because all that could be observed of creatures in a plane would be outline,—like the outline of a blot on a piece of paper. To view your neighbor, if you were Two-dimensional in expression, you would have to slide completely around him in your plane, and would know him by the shape thus described. Now think how tremendously you, a Three-dimensional creature (at present), could mystify plane-creatures; and you could most easily do so by using the One Two-dimensional form this world contains; One which is inseparable to your Three-dimensional existence,—a shadow! By casting your Two-dimensional shadow onto the plane, you would have "created," miraculously, an entity which, so far as Two-dimensional consciousness could discern, would be absolutely real, actual, conscious, living, vital, substantial, and intelligent. You could cause it to change its shape, grow large or small, and appear and disappear at will. Or supposing that you were so related to the plane, by proximity, that you could thrust your body through it at right angles,—through the third dimension, in other words. The performance would be almost identical with bobbing up through the plane-surface of a lake after a dive, with the exception that the movement would be continuous and even through the plane. First would come your head, a little circle gradually growing larger on the plane. Your movement would cause to appear in the plane an entity whose size and shape constantly changed due to the varying proportions of your body, although a knowledge of your Three-dimensional body would be impossible to the plane-creatures. Even the Number of forms produced by your body would vary as your arms, hands and fingers ascended through the plane. At last, as your toes emerged from the plane, the mysterious forms would disappear, leaving to the astonished Two-dimensional creatures no clue as to the law by which it was done.

Furthermore, imagine that you could communicate with the entities in a plane. From a Three-dimensional viewpoint you **Diagnosis** could look right through their bodies. If they were subject to ills, you could describe, as they might think "clairvoyantly," what the trouble was; and their own subsequent examination would prove your accuracy. Surely you must be a God to possess such knowledge!

By now you must have sensed the analogy toward which your thought is being directed. We have observed that a Three-**Shadows** dimensional entity casts a Two-di-**and** mensional shadow; then would not **Reality** a Four-dimensional entity cast a Three-dimensional shadow? Like your own shadow in a Two-dimensional plane, is it not reasonable to think that the Three-dimensional shadow would have every semblance of absolute reality to our Three-dimensional consciousness? In short, the shadow of a Three-or-more dimensional object, it seems logical to believe, would become a reality in the world of One less dimension than its own. Likewise the reality of any world of specified dimensions is evidently only a shadow of what is "real" in the next higher world. The reality of the Two-dimensional world having been discovered to be only a shadow in this world, may not our reality be but a shadow of what is real to the beings who live in Four dimensions? **And so may not all** dimensions be real only to consciousness? When conciousness shall have developed to a point of independence of all dimensions, and manifests as pure Spirit, **may not all forms of space, and all space itself, prove to be illusions?** These are some of the soul-stirring questions the Fourth Dimensional theory arouses.

As to the questions it answers, these, too, are many, and satisfying. We are now entering an age in which psychic phenomena, so called, are attracting tremendous attention. Yet despite the abundance of phenomena being produced by the spiritualists, despite the numerous books of "communications" being published, despite the efforts of the American Society for Psychical Research to classify the authentic data so abundantly in evidence,

Answered Queries

there is yet to be offered a satisfying, common-sense explanation of the basic law by which modern miracles are produced. That they are produced in accordance with law, science accepts; and scientists are beginning to discover some of the ways in which the law acts; but as to what the law is, they are at sea. "How," in this connection, is still an unexplained word. The Fourth Dimensional idea defines this "how". As to where we go after death, the answer is that we "go" nowhere, but simply change the focus of our consciousness from Three dimensions to Four. Spirit knows no dimensions, and only uses these for expression through form. Each birth and death represent a cycle in One form of expression. Knowing that the law of progress is imperious throughout nature, the logical thought is that death to this plane is birth upon a "higher" plane,—the **higher** refering specifically to dimensions of space. We do not necessarily leave this world, any more than a One-dimensional creature "leaves" his world, by the expansion of his consciousness from a string to a plane. He is still One-dimensional —plus; and may still function in his old world, in at least a limited degree. But his habitat becomes the plane-world, and to express in the old world to which he has "died" he must again focus his consciousness in One dimension.

So with us Three-dimensional beings. Death brings no change except a change of focus, and an environment of infinite-

After Death ly enlarged possibilities. To manifest to our loved ones who mourn us as dead, we would have merely to change our focus,—cast our shadow, a **Three-dimensional shadow** of our newly-evolved Four-dimensional consciousness, into the world to which it appears as reality. We would have to know the law to do so, and our attempts, until knowledge of the law was very general and firmly established in our consciousness, might be very inadequate; but the knowledge of the mere existence of such a law, in a way that makes it comprehensible to us, is a wonderful inspiration to us plodders of earth; and our comprehension is a prophecy of demonstration.

The Fourth dimensional world, then, is the astral world, which interpenetrates this in the same way that a cube can be said to include infinite Numbers of plane surfaces, or squares,—which represent, incidentally, in Two dimensions, what a cube does in Three.

The possibilities of original thought upon this subject are infinite. The idea is comparatively new to this age of civilization. It is in its infancy, but **Enlarging Our Cosmic Vocabulary** it will grow, and even the meagre outline of the idea presented here, fragmentary as it is, largely due to limited space for expression, will be sufficient, it is hoped, to enable the student to grow with the idea, to keep pace,—or even take a step in advance—of world progress, and add a few more "words" to his vocabulary in the Universal Language.

XXXIII. Mystic Numbers

WHILE this book is devoted chiefly to the consideration of the digitary Numbers (as being the basis for a consideration of all other Numbers) certain of the higher Numbers (quantitatively) seem to **"Higher"** have a significance peculiar to **Numbers** themselves and not directly derived from their digitary components. It can be readily seen why 100 and 1000 would symbolize an indefinite period of time required for the accomplishment of some particular task, since these are the cyclic Numbers; but the peculiar associations of ideas with the Numbers 666 and 888, for example, cannot be so lightly disposed of, and involve an elaborate system of what might well be termed "mysteries."

666 and 888 are Two of a series of Numbers representing different states of consciousness in the individual, and closely **Mystic** allied with Biblical teachings (par- **Numbers** ticularly as put forth in the Book of Revelation) but also taught in the Greek Mysteries of Initiation, and very probably in most secret orders of a spiritual nature as well.

Since the students this book is likely to reach, are most familiar with the Christian Bible, these Numbers, already re- **Basis of** ferred to in the foregoing Chapters, **Study** will be more thoroughly treated of

177

here in the light of the allusions and
veiled references to them in the Last Book
of the New Testament. In arriving at the
meaning of the original Greek and Hebrew
text, and to a considerable extent in interpreting the symbolical "characters" of this drama
of Numbers, the writer is indebted to James
Morgan Pryse, whose elaborate and exhaustive
volume, "The Restored New Testament" has
been very helpful.

There will be discovered in Revelation
(xii-xiii) the description of Four symbolical

**The Beast
In
Revelation**
Beasts: a Lamb, identified as
Jesus (Iesous) who was to lead
the 144,000 into salvation; a
Beast resembling a leopard, hav-
ing the feet of a bear and the
mouth of a lion, and owing his power to the
Dragon; the red Dragon with Seven heads and
Ten horns, his tail drawing after it a third
of the stars of heaven, 'the Devil' and the
'Adversary,' the ancient (evil) serpent, who
leads the whole **earth** astray; and a Beast with
two horns like a lamb, but speaking like a
Dragon. He causes the earth to worship the
first wild Beast, and works "great miracles,
even to make fire come down to earth."

"Here is wisdom. Let him that hath
understanding count the Number of the
beast, for it is the Number of a man, and
his Number is Six hundred Three score
and Six."

The word translated understanding is
"Nous," the familiar term in Greek philosophy

The Beasts In Man for the higher mind. This naturally suggests the Beast as being the lower mind, called the Phren (he phren). Numbers in Greek are expressed by letters instead of figures such as we use, and setting down the numerical equivalents for he phren, the result is 666. The first Beast (resembling a Leopard) is quite evidently referred to by this Number. The Four beasts correspond perfectly with Four divisions of the human body, generally recognized, and by reference to the Bible text with this idea in mind, the connection is made startlingly plain.

The Four divisions are, briefly; the Head, the Higher Mind, the Nous or Iesous (sometimes referred to as heaven, the **Four Parts of the Body** seat of man's highest intelligence and directive power,—the higher or impersonal mind) ; the Heart, the Lower or Phrenic Mind (the emotional center of the body, in which intelligence is influenced by the personal element) ; the Navel (or Bowels), Desire, Epithumia, (the seat of desire and sensation) ; and the Genitals, Sensuality, Akrasia (the generative center).

Since One of the beasts corresponds to One of these divisions, and is given as the "key to wisdom" by the Revel-**Comparison** ator, it is not straining common sense to relate the other Three beasts to the other divisions, and when they are compared there is found abundant corroboration for the idea.

Iesous, (the Higher Mind) in Greek enumeration becomes 888. The "fiery red Dragon," from its description, is

The Numbers of the Beasts

quite evidently "Desire," and works out to the Number 555. The fourth beast, leading the whole world astray, is even more evidently Sensuality. He is described as having power "to bring fire down to earth." Earth is the lower part of the body,—the generative center; the vital, spiritual essence of the body, (the Messiah) has often been described as a fire, and in the great majority of people is truly brought down to earth. It has the power to work miracles; which occultists have sometimes misused to produce miracles. Christ taught the spiritual use of this power; the Beast would have it used carnally. "Akrashia," Sensuality, works out 333.

From the above the following incomplete Table is directly obtained. It indicates unmistakably the purpose and intent of the Revelator in teaching the spiritual growth of the individual.

Meaning of the Diagram

Following One Number to the next above is really tracing the "level of consciousness" in man, as he gradually evolves an increasingly spiritual concept of life. It is a wonderful thing to grasp the thought that there are correspondences on all planes of being. Our mental and spiritual growth is paralleled by a change in bodily forces,—or, more particularly,—in their use.

You and the Universe, a Book of Numbers

Head	888 Iesous, the Higher Mind,	I. "The Lamb"
Heart	666 He Phren, the Lower Mind,	II. "The Beast"
Navel	555 Epithumia, Desire,	III. "The Red Dragon"
Genitals	333 Akrasia, Sensuality or Akolasia, Licentiousness	IV. "The Adversary"

It will be seen that in the numerical progression from 333 to 888 there are two missing numbers, 444 and 777, and above **The Good** 888, we might add 999 and 1000. **Serpent** 333 has been called the evil serpent, and would suggest a "good" serpent. This would correspond to the wise use of generation (instead of its mis-use) and would lead to regeneration. This regenerating process, by which the desires are spiritualized, is a gradual one, and has been referred to mystically as the "coil of the serpent." The course of the electric or vital principle in the body, if not destroyed by the Adversary, is that of a spiral, or coil, like part of a figure 8. It is described by the Greek word speirema, whose value is 444. When the lower mind of emotion and a personal viewpoint of life is displaced by the larger concept **The Cross** of the higher mind, which sees life in the universal, impersonal

Christ-sense, the "level of consciousness" in body (carrying out the idea of correspondence) actually "crosses" from the upper part of the trunk to the head, at "the place of the skull" (Golgotha in the Bible actually means this). Thus it is crucified or "cross-i-fied" that the world of flesh might be redeemed or renewed by spirit, or the spiritual force. The Greek word Stauros, meaning "cross" and with particular reference to the cross of Christ, fills our requirements and numbers 777.

The Light of the World

The action of the spiritual force upon the brain is like that of electricity (perhaps it is electricity) and greatly increases its powers. It radiates in Three directions from the base of the brain, forming a cross. It, so to speak, floods the brain with light, or illumines it. ("I am the light of the world"). Such a development ("the raising of the serpent in the wilderness") gives the individual true psychic perception, as contrasted to the misuse of the same power on the sensual plane, and results in the highest powers, so it is declared, that man can attain in the body. It is described by the Greek word episteme, whose numerical equivalent is 999, and means "The Intuitively Wise." Truly he who thus achieves dominion over "the beasts" and "every living thing" is a conqueror, what the Greeks would describe by ho nikon,—and this expression adds 1000. So the wonderful drama of the Revelation is expressed; the sublime epic of man's victory over self, the never-ending symphony of attainment.

Herewith is presented the complete table or chart, (somewhat after that presented by Pryse and others,—for the truth is One, and as men grow in understanding of it, their conclusions or presentations grow increasingly alike). The present writer has changed some terms for readier understanding, has added to the fullness of the theme, and endeavored to simplify its forms as much as might be:

Head	1000	Ho Nikon, "The Conqueror"
(Spirituality)	999	Epistemon, Intuitively Wise, the True Seer
	888	Iesous, the Higher Mind, the Impersonal Sonsciousness
		I. THE LAMB
	777	Stauros, the Cross
Heart	666	He Phren, the Lower Mind, the Personal Consciousness
(Emotion)		II. THE BEAST
Navel	555	Epithumia, Desire, Sense Consciousness
(Sensation)		III. THE RED DRAGON
Genitals	444	Speirema, the Serpent Coil
(Generation)	333	Akrasia, the False Seer, Carnal Consciousness
		IV. THE ADVERSARY

XXXIV. Color Analogies

COLORS were One of the first means of expressing religious ideas and of transmitting thought and preserving memory. In the primitive religions of Mexico and Egypt colors were used to express definite religious ideas until, with the grad-

Religion Color in
ual substitution of the symbol for the thing it represented, the religious meaning was lost and color combination became an art of pleasing the sense of sight. The spiritual meaning of colors was subordinated to the material or objective appeal of the eye. Archaeologists, in unearthing very ancient Indian and Egyptian paintings, remarked upon the total absence of tints, and the use of only solid colors, little guessing that the hidden spiritual significance was of far greater concern to the painter than the accident of whether the use of clear colors formed a pleasing visual combination.

At first color was used to express only spiritual ideas; regardless of physical appeal it spoke the same clear language

The Evolution of Painting
to all. Then the painters sought to veil the spiritual message by compromising and appealing to the eye as well, and with the degeneration of religion, color symbolism fell into disuse in relation to it and became a device

of heraldry. Semblances of the symbolism were carried out on the coats of arms of the nobility; but finally even this use became legendary and the application of color became a material art whose spiritual possibilities were known only to the few.

Legend and Folk-Lore have attributed to the several colors certain definite influences which the layman must either **Legend and** accept or reject as his reason **Folk-Lore** and intuition prompt him, for without the hidden key to this mystery which the Science of Numbers gives, there is no very definite way of verifying the truth of such statements. They deal with spiritual qualities which are not always easily detected objectively, although sufficient proof of their accuracy has accumulated to maintain their association throughout the ages; and the discovery of an inter-relationship between the "octaves of vibration" and thought is confirming these old beliefs.

Every Color has a Two-fold spiritual meaning: a good and an evil. In its favorable aspect red, for instance, is the **Dual** symbol of physical life, strength, vig- **Meaning** or, courage, victory. It is the emblem of bravery as used in our national flag. There is nothing more conducive to awakening the feeling of good will, fraternity and good fellowship than to sit about a great fireplace from which the red embers of the fire cast their warming light over the assembled company. Such a light is referred to as a "rosy" light, and rose is the spiritual aspect of the color red. Crimson might be called its physical expression as it pulses through the veins of all humanity. It was the rose-color that Christ is depicted as having worn during his ministry, symbolizing the great impersonal, spiritual love he felt for all life.

The Seven-fold Path

Red vibrates the lowest in the scale of colors and is comparable to the tonic in the musical scale. Rose is its eighth or octave note, the same principle evolved upon a higher plane. The Seven colors of the spectrum compare with the Seven great religions, the Seven planets, the Seven days of the week, the days of creation, the Seven initiations, the virtues and vices, the candlesticks, stars, churches, seals, angels, and trumpets of Revelation, the Seven principles in man, and the Seven Hermetic principles of Mentalism. Man apparently comes under the influence of each of these colors at some period in his life, and exteriorally this fact is shown in birth and death. The tiny new born babe is a living mass of red, and when the Seven-fold path of life has been traveled and Death approaches, he places a violet cast over the silent form in token that the journey has been completed.

Negative Red

The negative, destructive, evil aspect of red has gained more prominence in human thought than has the desirable side, and this negative expression is seen in its effect upon certain animals and birds, such as the bull and the turkey gobler. When persons are angry their faces flush and they "see red" or "see blood." Destructively red is the symbol of unrest, violent emotion, discontent, destruction (as when property is razed by fire); and the factions of unrest in modern civilization have adopted the red flag. On the other hand is the Red Cross, —the symbol of Mercy and Brotherhood.

Orange Orange is the symbol of Whole-ness, formed by the blending of red (life) and yellow (wisdom). It augurs well for prosperity, self-control, and soundness. Moderation should be its motto. Orange possesses a strong healing force; it strengthens the powers of elimination (many medicines used for this purpose are orange in their hue). In its negative sense it develops the type of person who is morbid on the subject of health; extreme physical culturists, the food faddists, the professional athlete, the mania of hypochondriasis.

Yellow Yellow is the symbol of wisdom, knowledge, illumination, aspiration, justice. It is the priceless gold referred to in the Bible, the honey which is so often referred to. It teaches the lesson that wisdom is only truly such when it is acquired and used in Righteousness; it is gained with effort, just as the precious yellow-gold metal is extracted from the great quantities of base metal and ore. It must be of such purity as to bear unflinchingly the bright yellow sunlight of Righteousness or right use.

188

You and the Universe, a Book of Numbers

We read in the old proverb "A little knowledge is a dangerous thing." A lamp in which the golden oil burns low is an unsafe guide. The Foolish Virgins had but a little of this oil of wisdom in their lamps. A shallow, foolish person parades the little wisdom he has until it is reduced to the plane of the ridiculous. The wise let their light so shine as to proclaim itself; true wisdom needs no advertisement. "Silence is golden."

The Foolish Virgins

Sometimes a little wisdom is the fore-runner of jealousy,—that terrible, destructive aspect of yellow. Sometimes, too, it manifests in deceit. Judas has been portrayed by the painters of old as wearing a robe of dingy yellow—wisdom tinged with the blackness of deceit.

Negative Yellow

Green is the color of energy and rest. The ability truly to rest implies the strength really to be active. Green is the color of individuality, of fruitfulness, of renewal and supply. If you are depleted stretch out at full length upon the greensward and let the bountiful supply of nature's predominant color renew you. Green is likewise a symbol of prosperity in the sense of a never-failing supply; it will not admit of extravagant wastefulness or of idleness (recall the fable of the grasshoppers and the ants), but always maintains a sufficiency when that sufficiency is earned. It teaches that "what you give, you have; what you keep, you lose." Plants re-

Green

188

main green so long as they radiate their beauty
in the sunlight; shut them up in a closed room,
attempt to narrow their sphere of usefulness,
and their bright greenness leaves them and
they are robbed of color. Green is the color
of reciprocity, hope, and youth. We speak of
a green sapling, and of a green old age. When
the Psalmist rested in green pastures he said,
"Thou restoreth my soul."

Negative green symbolizes avarice, jealousy and envy. "Turned green with envy" is **Negative Green** an expression giving voice to this esoteric truth. Artists depict misers in a green light, their features overcast with a cold blue-green hue.
The theatre resorts to the same device, "to get
its message across" by means of color.

Blue is the symbol of truth and purity and
gentleness; of reality, faith, innocence, beauty.
Blue The Virgin is depicted in a light blue robe (gentleness). The name Mary is from a root word Mar-e, meaning water,—the symbol of purity. We speak of
people being "true blue." Blue garments seldom fade and blue is so dependable a color
that dyers use it as the basis of their colors
whenever possible.

Blue, like truth, is cold unless warmed by
the touch of the rose-color of life. It can be
Negative Blue cruel and keen and sharp of itself—like cold blue steel; a rather repellant attribute. Truth should be spoken only in love if One would
manifest its positive, constructive side. Rooms

decorated all in blue, unless the blue is tinged with the warm yellow of wisdom or the rose-light of love, are cold and unpleasant. A bright sunny room tinted in blue is attractive because truth is then made cheery by the sunlight of wisdom and mercy.

      Purple is the color of power. We speak of "royal purple". Men of great ungoverned power actually become purple in the **Purple** face. Veins enlarged and filled with impure blood, stand out upon their features. To be desirable, might must be tinged with humility. The really great man is the humble man. Crude, glaring purple, composed of the crimson of physical love, and the cold, cruel blue of truth, manifests a color which symbolizes the passions,—a dangerous, ominous color. It must combine the higher development of both red and blue to be helpful. Then it is truly royal and the fit habiliment of kings, kings who have dominion over the most unruly of kingdoms,—self.

      When this refinement takes place purple begins to merge into violet; the beautiful blending of light-blue (gentleness) and rose-**Violet** color (love). Then it becomes the color of God's messengers, who are the interpreters of His word, the friends and elder brothers of humanity; a color of the meek and lowly, and such the truly consecrated are.

      Finally when life's lessons have been learned, when the Seven-fold path of attain-

You and the Universe, a Book of Numbers

The Light of the World ment has been travelled, all of the colors merge back into the unity of pure crystalline white. Such was the color of the robe in which Christ appeared after the resurrection. The red tinge of physical life was gone and the unmixed whiteness of spiritual understanding and At-one-ment had been attained. The white light is made up of all the colors. It is the light of the world; such was the Christ the Son (Sun) of God. The purity of the Christ and the whiteness of sunlight symbolize One thing. We speak of crystalline qualities and we well know that they are Christ-like. "As spiritualists gaze into the crystal in order to read the future so may we as spiritual beings gaze at Christ in whom is revealed the future of our being, the ultimate perfection of the Sons of God."

The Ark of the Covenant is but a variation of the rainbow promise which was the Arc of **The Ark of the Covenant** the Covenant. As the Seven colors of the rainbow formed such an arc in the exterior world of the heavens, so in the heaven within us they find similar expression and have a correspondence in the Seven principles. Christ, the perfect type-man, fulfilled the promise of the arc, reflecting these principles, attributes, or colors within his spiritual nature, thus completing the Arc (an incomplete part of the circle of Being) by forming a circle which is the union of the exterior world and the interior world. He called the "Son,"—

cognate with "Sun" whose light contains Seven Colors, another hint at the true nature of His mission on earth,—to arouse mankind to a consciousness of their Godhood or Divine Oneness. He said, "I AM the light of the world," "Ye are the light of the world;" but few have ever guessed that He meant an inner light, corresponding to the sunlight outside.

The story of the Covenant in Genesis means much to him who knows that the Seven colors combined as in the Rainbow, are a symbol of regeneration, **The Rainbow** of spiritual rebirth or resurrection. And since the rainbow is but white light separated we can understand why Christ, after his Resurrection or reunion with God, is depicted as wearing snowy white.

In Mythology, Iris, the messenger of the gods, whose symbol was a rainbow, was merely another beautiful device of pre- **Mythology** senting the truth of man's rebirth in spirit. In Egypt, the goddess Isis is always represented in a robe in which all the colors are blended, symbolizing her possession of all attributes. Isis means light, and she was the Goddess of Nature. Osiris, represented as wearing a white headdress, is another expression of the symbolism of color; the difference between the significance of the rainbow and white being that the latter represents spiritual understanding—the inner attribute,—while white separated into its component colors symbolizes God in Nature, the outer manifestation.

You and the Universe, a Book of Numbers

References to the symbolism of white as well as to other colors are very numerous in

Color in Literature

the Bible; but like all things, unless our attention is called to it, we are not likely to observe it. The Bible was written by men skilled in the occult sciences, students of Astrology, the Science of Numbers and Symbolism. When their writings are interpreted literally much of the wonderful beauty they contain is lost, just as we can see no beauty or religion in the sacrificial rites of the Egyptians, because we see the expression only and not the spirit. Jehovah ordered Aaron not to enter the s tuary until he had clothed himself in white linen, which is to say,—"not until he had developed spiritual understanding." The Magi of the East always wore white. In many countries where people recognize and reverence symbolism more than we, white cows and white elephants are considered sacred. This is notable in India and some parts of China. In Persia the white horse was considered sacred and was consulted as an oracle in crises. The reverence was due to the original wonderful meaning of the color white. The present veneration of these animals has retrograded into mere idolatry and worship; and probably the worshippers no longer know the hidden reason why this custom of veneration has come down to them.

Mythology always represents the gods who are beneficent as being drawn by white horses; those of evil intent by

White and black. The Science of Numbers
　Black　tells us that the objective meaning of black and white are the same, which again reminds us of the Law of Opposites. In the Language of Symbolism each color is interpreted as good or evil by its combination with other colors. The spiritual influence of white was recognized by Pythagoras and he even insisted on the hymns being sung in the temples by white-robed choristers. White has always been associated with spiritual understanding, regeneration, illumination, and the conference of blessings and favors. White swans are often bearers of gifts in Mythology. Wagner has Parsifal drawn by white swans in his opera of that name. Black swans are, of course, the omen of evil. Another very striking clue to the meaning of white has survived even in our time. Observe, for instance, the relation of the word "white" and "wit." Wit is now narrowed in its meaning, but still bears some resemblance to its older significance, which was understanding or comprehension. In the German the present tense of the verb "to know,"—"wissen,"—is the same as the word meaning white; thus "ich weiss" meaning "I know" and "weiss,"— "white."

　　This Lesson would be incomplete without just passing reference to the very striking analogy observed in the Incorporated Seal of The Harmonial Institute
　The　For Re-Education. The prominence
Institute　in its arrangement is a Triangle.
　Seal

point upward, whose sides are inscribed LOVE, WISDOM, and TRUTH. The Colors which correspond to these attributes are Red, Yellow and Blue,—the Three primary colors from which all Seven in the spectrum are evolved. It will be of interest to the student to know that when the Seal was designed this color analogy was not observed and the striking manner in which the arrangement is seen to correspond with various forms of symbolism is a suggestion of the high inspiration which is back of The Harmonial Philosophy.

> I think today is sparkling red,
> The vital, living, manifest;
> And yesterday, it might be said,
> Is yellow, wisdom guiding best.
> For hope and love, tomorrow's blue;
> In all are blended truth and light.
> As in the colors, running through,
> We see the One, the perfect white.
> —John Willis Ring.

NUMERICAL CORRESPONDENCES

Gems	Colors	Flowers
1. Aquamarine Turquoise	1. Flame Lilac Crimson	1. Lilac
2. (See 11)	2. Gold	2. (See 11)
3. Amethyst Ruby Amber Sardonyx	3. Gold-flame Slate Rose-red	3. Rose Orchid Pansy Forget-me-not Mignonette
4. Emerald Moonstone Bloodstone	4. Blue Green Terra Cotta Indigo	4. Columbine Fuchsia Goldenrod
5. Coral	5. Pink	5. Sweet Peas Primrose Carnation
6. Diamond Topaz Jasper Onyx	6. Orange Heliotrope Scarlet	6. Heliotrope Tuberose Chrysanthemum
7. Agate	7. Steel Purple Brick	7. Hyacinth Geranium Poppy
8. Opal Pearl	8. Canary Tan Bronze	8. Jonquil Rhododendron
9. Malachite	9. Red Brown Lavender	9. Aster
10. (See 1)	10. (See 1)	10. (See 1)
11. Sapphire Garnet Jade	11. White Yellow Violet Black	11. Violet
22. (See 4)	22. Cream	22. Daisy

www.ingramcontent.com/pod-product-compliance
Lightning Source LLC
Chambersburg PA
CBHW051908170526
45168CB00001B/299

* 9 7 8 1 5 1 5 0 5 6 5 0 8 *